INDUSTRIAL STATISTICS

INDUSTRIAL STATISTICS
Practical Methods and Guidance for Improved Performance

ANAND M. JOGLEKAR
Joglekar Associates
Plymouth, Minnesota

WILEY

A JOHN WILEY & SONS, INC., PUBLICATION

Library of Congress Cataloging-in-Publication Data:
Joglekar, Anand M.
  Industrial statistics : practical methods and guidance for improved
performance / Anand M. Joglekar.
        p. cm.
  Includes bibliography references and index.
  ISBN 978-0-470-49716-6 (cloth)
1. Process control–Satistical methods. 2.   Quality control–Statistical
methods. 3. Experimental design. I.   Title.
  TS156.8.J62 2010
  658.5072'7–dc22

                                              2009034001

To the memory of my parents
and to Chhaya and Arvind

The following age-old advice deals with robust design and continuous improvement at the personal level.

You have control over your actions, but not on their fruits.
You should never engage in action for the sake of reward,
nor should you long for inaction.
Perform actions in this world abandoning attachments
and alike in success or failure,
for yoga is perfect evenness of mind.

– Bhagavad Gita 2.47–48

Mahatma Gandhi encapsulates the central message of Gita in one phrase: *nishkama karma*, selfless action, work free from selfish desires. Desire is the fuel of life; without desire nothing can be achieved. *Kama*, in this context, is selfish desire, the compulsive craving for personal satisfaction at any cost. *Nishkama* is selfless desire. *Karma* means action. Gita counsels—work hard in the world without any selfish attachment and with evenness of mind.

Mahatma Gandhi explains—By detachment I mean that you must not worry whether the desired result follows from your action or not, so long as your motive is pure, your means correct. It means that things will come right in the end if you take care of the means. But renunciation of fruit in no way means indifference to results. In regard to every action one must know the result that is expected to follow, the means thereto and the capacity for it. He who, being so equipped, is without selfish desire for the result and is yet wholly engrossed in the due fulfillment of the task before him, is said to have renounced the fruits of his action. Only a person who is utterly detached and utterly dedicated is free to enjoy life. Renounce and enjoy!

– Adapted from Bhagavad Gita by Eknath Easwaran

CONTENTS

PREFACE

This book is based upon over 25 years of teaching and consulting experience implementing statistical methods in a large number of companies in industries as diverse as automotive, biotechnology, computer, chemical, defense, food, medical device, packaging, pharmaceutical, and semiconductor among many others. The consulting assignments have resulted in many success stories—large cost reductions, rapid product development, regaining lost markets, dramatic reductions in variability, and troubleshooting manufacturing. Over ten thousand participants have attended my seminars on statistical methods. All these interactions—the technical problems the participants brought forward, the prior statistical knowledge they had, and the questions they asked—have shaped the writing of this book.

Much of the technical work in industry relies upon the coupling of known scientific and engineering knowledge with new knowledge gained through active experiments and passive observations. Accelerating this data-based learning process to develop high-quality, low-cost products and bringing products to market rapidly are key objectives in industry. The fact that statistical methods are a necessary ingredient in accomplishing these objectives is as true today, if not more so, as it was 25 years ago. The use of statistical methods by all technical individuals in industry, which number in the millions, continues to be an important need.

Four major changes have occurred during the past several years that have influenced the writing of this book:

1. With the advent of personal computers and statistical software, the need to understand statistical computations, in the detail necessary for hand calculations, has reduced dramatically. Today, the job of number crunching can and is delegated to a software package. The statistical computations described in great

detail in various textbooks on statistics are interesting to know but their mastery is no longer necessary to make good applications. This means that the focus has to be on explaining concepts and logic, practical guidance on the correct use of statistical methods, interpretation of results, and examples to demonstrate how to use the methods effectively.

2. As a result of the various iterations of quality approaches—TQM, BPM, Process Reengineering, Six Sigma—there is a greater awareness and focus on the use of statistical methods even in industries where such use was almost nonexistent a short time ago. People are more familiar with statistical methods than they were years ago. This means that a certain degree of statistical knowledge on the part of the audience can be presumed. I have based the knowledge that can be presumed on my experience with the audience.

3. International competition and the need for much higher productivity have resulted in increased workload for technical individuals. There is less time to do more work in! This means that information needs to be presented compactly and in a focused manner dealing with only those issues that are of the highest practical importance. The book needs to be concise and to the point.

4. Managers and black belts are now responsible to promote and implement statistical methods in a company, a job that previously was being done almost exclusively by statisticians. Managers and black belts have various degrees of statistics knowledge but they are not full-fledged statisticians. They need help to implement statistical methods. The book needs to include guidance on implementation.

This book is specially written for the technical professionals in all industries. This audience includes scientists, engineers, and other technical personnel in R&D and manufacturing, quality professionals, analytical chemists, and technical managers in industry—supervisors, managers, directors, vice presidents, and other technical leaders. Most of this audience is engaged in research, product design, process design, and manufacturing, either directly or in support roles. A significant portion of their job is to make decisions based upon data. To do this well, they need to understand and use the statistical methods. This book provides them with the main concepts behind each of the selected statistical method, examples of how to use these methods, and practical guidance on how to correctly implement the methods. It also includes an extensive chapter on questions and answers for the reader to practice with. The material is presented in a compact, easy-to-read format, minimizing the mathematical details that can be delegated to a computer unless mathematical presentation illuminates the concepts. Most of this audience has access to some statistical software package (software 2009). Many are not interested in the details of statistical computations. For those who are so inclined, this book provides recommendations for further reading. Many in this audience such as technical managers, technical leaders, and black belts also have the responsibility to help guide the implementation of the statistical methods. This book identifies questions they should ask to help accomplish this objective.

This book concisely communicates 10 practically useful statistical methods widely applicable to research, product design, process design, validation, manufacturing, and continuous improvement in many different industries. The following criteria were used to select the statistical methods and particularly, the emphasis placed on them in this book.

1. The selected method is widely applicable in R&D and manufacturing in many industries.
2. The method is underutilized in industry and wider use will lead to beneficial results.
3. The method is being wrongly used, or wrong methods are being used, to solve the practical problems at hand.
4. There are misconceptions regarding the method being used that need to be clarified.

ORGANIZATION OF THE BOOK

This book contains 11 chapters. The last chapter includes a test (100 practical questions) and answers to the test. People familiar with the subject matter may take the test and then decide what to focus on, whereas others may read the book first and then take the test. Brief outlines of the remaining 10 chapters follow:

1. *Basic Statistics: How to Reduce Long-Term Portfolio Risk?* This chapter introduces the basic statistical concepts of everyday use in industry. These concepts are also necessary to understand practical statistical methods described in the remaining chapters of this book. Most people in industry, including those who have just joined, are interested in investing their 401k contributions in stocks, bonds, and other financial instruments to earn high returns at low risk. This question of portfolio management, which formed the basis of the Nobel Prize-winning work of Prof. Markowitz on mean–variance optimization, is used as a backdrop to explain the basic statistical concepts such as mean, variance, standard deviation, distributions, tolerance intervals, confidence intervals, correlation and regression. The properties of variance, and in particular, how risk reduction occurs by combining different asset classes are explained. The chapter ends with questions to ask to help improve the use of basic statistics.

2. *Why Not to Use a t-Test and What to Replace It With?* It was almost exactly 100 years ago that the *t*-distribution and the *t*-test were invented by W. S. Gosset. This important development provided the statistical basis to analyze small sample data. One application of a *t*-test in industry today is to test the hypothesis that two population means are equal. Decisions are often made purely based upon whether the difference is signaled as statistically significant by the *t*-test or not. Such an application of the *t*-test to industrial practical

problems has two bad consequences: practically unimportant differences in mean may be identified as statistically significant and potentially important differences may be identified as statistically insignificant. This chapter shows that these difficulties cannot be completely overcome by conducting another type of *t*-test, by computing sample sizes, or by conducting postexperiment power computations. For practical decision-making, replacing the *t*-test by a confidence interval for difference of means resolves these difficulties. Similar arguments apply to all other common hypothesis tests, such as the paired *t*-test and the *F*-test. Many practical applications are considered throughout the chapter. The chapter ends with questions to ask to help improve data-based decision making.

3. *Design of Experiments: Is It Not Going to Cost Too Much and Take Too Long?* In industry, there continues to be insufficient understanding and applications of the important subject of design of experiments. There is also a misconception that designed experiments take too long and cost too much. This chapter shows how, through efficient and effective experimentation, designed experiments accelerate learning and thereby accelerate research and development of products and processes. It illustrates the many pitfalls of the commonly used one-factor-at-a-time approach. It explains the key concepts necessary to design, analyze, and interpret screening and optimization experiments. It identifies the considerations that must be well thought through for successful applications of the design of experiments. Many practical applications are considered throughout the chapter. The chapter ends with questions to ask to help implement and improve the use of designed experiments.

4. *What Is the Key to Designing Robust Products and Processes?* Robust design method needs to be more widely understood and implemented. It adds two important dimensions to the classical design of experiments approach. The first important dimension is an explicit consideration of noise factors that cause variability and ways to design products and processes to counteract the effects of these noise factors. This chapter explains the basic principle of achieving robustness. Robust design means reducing the effect of noise factors by the proper selection of the levels of control factors. Robustness can be achieved only if control and noise factors interact. This interaction is the key to robustness. The design and analysis of robustness experiments is illustrated by examples. The second important dimension of robust design is a way to improve product transition from bench scale research to customer usage such that a design that is optimal at the bench scale is also optimal in manufacturing and in customer usage. Knowledge gained at the laboratory stage does not easily transfer during scale-up because of control factor interactions. The ways to reduce control factor interactions are explained. These two new dimensions have major implications toward how R&D should be conducted. Many practical applications are considered throughout the chapter. The chapter ends with questions to ask to help implement and improve the use of the robust design method.

5. *Setting Specifications: Arbitrary or Is There a Method to It?* Specifications for product, process, and raw material characteristics are often poorly set in industry. This chapter begins with the meaning of specifications and the implications of predefined specifications toward variability targets that must be met in R&D and manufacturing. The basic principles of setting specifications using three different approaches are explained with several examples. The three approaches are empirical approach, functional approach, and minimum life cycle cost approach. The functional approach includes worst case, statistical, and unified specifications. Many practical applications are considered throughout the chapter. The chapter ends with questions to ask to help improve specification development.

6. *How to Design Practical Acceptance Sampling Plans and Process Validation Studies?* The design of acceptance sampling plans and process validation studies is inadequately done in industry. This chapter clarifies the misconceptions that exist in industry regarding the protection provided by the sampling plan. These misconceptions occur because insufficient emphasis is placed on understanding the operating characteristic curve of a sampling plan. Once the acceptable quality level (AQL) and the rejectable quality level (RQL) are selected, the software packages instantly design the sampling plans. The chapter provides practical guidance on how to select AQL and RQL. It explains the connection between AQL and RQL to be used for process validation and the AQL and RQL to be used in manufacturing for lot acceptance. Often, validation studies are designed with inadequate sample sizes because this connection is not understood. Many practical applications are considered throughout the chapter. The chapter ends with questions to ask to help improve the design of validation studies and acceptance sampling plans.

7. *Managing and Improving Processes: How to Use an At-A-Glance-Display?* Statistical process control is widely used in industry. However, control charts are indiscriminately used in some companies without realizing that they are useful only if the process exhibits certain behavior. Control charts are often implemented without an adequate consideration of the risk and cost implications of the selected chart parameters. Also, quality reviews are often inefficient and ineffective. This chapter explains the fundamental rationale behind the development of control charts. It provides practical guidance to select subgroup size, control limits, and sampling interval. And it provides an at-a-glance-display of capability and performance indices making it easier to plan, monitor, review, and manage process improvements. Many practical applications are considered throughout the chapter. The chapter ends with questions to ask to improve process management.

8. *How to Find Causes of Variability by Just Looking Systematically?* This chapter deals with the much underutilized topic of variance component analysis. Reducing variability is an important objective in manufacturing. Variance components analysis helps identify the key causes of variation and the contribution of each cause to the total variance. This chapter explains the basic principles

of variance components analysis, how such an analysis can be done with data routinely collected in manufacturing, and how the results can be used to develop cost-effective improvement strategies. The principles of designing variance component studies, including the appropriate selection of the degrees of freedom for each variance component, are explained. Many practical applications are considered throughout the chapter. The chapter ends with questions to ask to help find causes of variability and make cost-effective improvement decisions.

9. *Is My Measurement System Acceptable and How to Design, Validate, and Improve It?* Some key questions often asked in industry are: How to know if the measurement system is adequate for the job? How to design a robust measurement system? How to demonstrate that the measurement system is acceptable, and if not, how to improve it? This chapter provides the acceptance criteria for measurement system precision and accuracy, for both nondestructive and destructive measurements. The rationale for the acceptance criteria is explained. The principles of designing cost-effective sampling schemes are explained. An example is presented to show how robust product design ideas can be used to design a robust measurement system. A design of experiments application is considered to demonstrate how to cost-effectively validate a measurement system and how to develop specifications for measurement system parameters. A gage repeatability and reproducibility application is considered to demonstrate how the acceptability of the measurement system can be assessed and how the measurement system can be improved if necessary. Many practical applications are considered throughout the chapter. The chapter ends with questions to ask to design and improve measurement systems.

10. *How to Use Theory Effectively?* While technical professionals learn a great deal of theory during their undergraduate and graduate education, theory is often not extensively and effectively used, perhaps because it is felt that theory does not work perfectly in practice. There is much to be gained, however, by the judicious combination of theory and data. The purpose of this chapter is to introduce the subject of model building, both empirical modeling based purely upon data and mechanistic modeling based upon an understanding of the underlying mechanism. A theoretical equation for coat weight variability of controlled release tablets is derived to demonstrate how mechanistic models can be built. The equation permits the coating process settings to be optimized without much experimentation. Many practical applications are considered throughout the chapter. The chapter ends with questions to ask to help put greater emphasis on the use of theoretical knowledge coupled with data.

HOW TO USE THIS BOOK

This book can be used in many ways. It can be used for self-study. It can be used as a reference book to look up a formula or a table, or to review a specific statistical method.

It can also be used as a text for quality-statistics courses or engineering-statistics courses for seniors or first-year graduate students at various universities. It should help provide university students with a much-needed connection between statistical methods and real world applications.

The topics in the book are generally arranged from those most useful in R&D to those most useful in manufacturing. Readers who wish to study on their own should first review the table of contents, decide whether they are or are not generally familiar with the subjects covered in the book, and then take the appropriate one of the following two approaches.

For those generally not familiar with the subject

1. Start reading the book from the front to the back. Go over a whole chapter keeping track of topics that are not clear at first reading.
2. Read through the chapter again, paying greater attention to topics that were unclear. Wherever possible, try to solve the examples in the chapter manually and prove the concepts independently. Note down the key points learned from the chapter.
3. If you feel that you have generally understood the chapter, go to the last chapter that contains test questions. These questions are arranged in the order of the chapters. Solve the questions pertaining to your chapter. Compare your answers and reasons to those given in the answer section of the last chapter. If there are mistakes, review those sections of the book again.
4. Obtain an appropriate software package, type in the data from various examples and case studies given in the book, and ensure that you know how to get the answers using the software of your choice.
5. Think about how these statistical methods could be applied to your company's problems. You are bound to find applications. Either find existing data concerning these applications or collect new data. Make applications.
6. Review your applications with others who may be more knowledgeable. Making immediate applications of what you have learned is the key to retain the learning.

For those generally familiar with the subject

1. Start by taking the test in the last chapter. Take your time. Write down your answers along with the rationale. Compare your answers and rationale to that given in the last chapter. Circle the wrong answers.
2. Based upon the above assessment, identify the chapters and sections you need to study. For these chapters and sections, follow the six steps outlined above.

There are many books written on the material covered in each chapter of this book. I have recommended appropriate books for further reading should additional information become necessary. Most of these books focus on one or two topics in considerable detail. Also recommended is my previous book *Statistical Methods for Six Sigma in R&D and Manufacturing*, published by Wiley in 2003, which should be treated as a companion book to the present offering.

I hope that you, the reader, will find this book helpful. If you have suggestions and comments, you can reach me at www.JoglekarAssociates.com.

ANAND M. JOGLEKAR

Plymouth, Minnesota

CHAPTER 1

BASIC STATISTICS: HOW TO REDUCE FINANCIAL RISK?

This chapter introduces the basic statistical concepts of everyday use in industry. These concepts are necessary to understand the practical statistical methods described in the remaining chapters of this book. Many people in industry are interested in investing their 401k contributions in stocks, bonds, and other financial instruments to achieve high returns at low risk. This question of portfolio management is used as a backdrop to explain the basic statistical concepts of mean, variance, standard deviation, distributions, tolerance intervals, distribution of average, confidence intervals, sample sizes, correlation coefficients, and regression. The properties of variance and how risk reduction occurs by combining different types of assets are explained.

The 1990 Nobel Prize in economics went to Professors Markowitz, Miller, and Sharpe for their contributions to the field of portfolio management. Professor Markowitz, who originated the subject, describes the beginning of this approach:

> The basic concept of portfolio theory came to me one afternoon in the early 1950s in the library while reading Williams's *Theory of Investment Value*. Williams proposed that the value of a stock should equal the present value of its future dividends. Since the future dividends are uncertain, I interpreted the proposal to mean the *expected* value of future dividends. But if the investors were only interested in the expected returns of the portfolio, to maximize the return they need only invest in a single security. This I knew was not the way investors acted. Investors diversify because they are concerned with risk as well as return. Variance came to mind as a measure of risk. The fact that portfolio variance depended on security covariances added to the plausibility of the approach.

Industrial Statistics: Practical Methods and Guidance for Improved Performance   By Anand M. Joglekar
Copyright © 2010 John Wiley & Sons, Inc.

Since there were two criteria, risk and return, it was natural to assume that investors selected from the set of optimal risk–return combinations.

With investments in 401k plans and elsewhere, many are interested in managing investments in stocks, bonds, and other financial instruments to achieve satisfactory returns at low risk. Given the precipitous drop in the stock market at the time of this writing, the subject of portfolio management is uppermost in the minds of many individuals. The debate over "time in the market" and "timing the market" has ignited. The purpose of this chapter is to communicate the basic concepts of statistics: mean, variance, standard deviation, distributions, tolerance intervals, distribution of average, confidence intervals, sample sizes, correlations, and regression. These concepts are used extensively in industry and are necessary to understand the statistical methods presented in the rest of this book. I have used the portfolio allocation problem with a long time horizon as a backdrop to explain basic statistics, not to suggest how you should allocate your money. Along the way, we will see how to play with means and variances in an attempt to increase long-term portfolio return while reducing risk. A more conventional and detailed discussion of basic statistics is provided in Joglekar (2003).

1.1 CAPITAL MARKET RETURNS

Table 1.1 shows the annual returns for large company stocks, small company stocks, international stocks, corporate bonds, and treasury bills over a period of 80 years from 1926 to 2005. Also shown are the changes in inflation as measured by the consumer price index (CPI). The data are as reported in Ibbotson (2006).

Large Company Stocks: The returns are for the S&P 500 or an equivalent index representing the average performance of large company stocks. There is considerable variability in returns: from an exhilarating 53.99 percent in 1933 to a depressing −43.34 percent in 1931.

Small Company Stocks: The returns are for the index of small company stocks. There is an even greater variability in returns: from a high of 142.87 percent in 1933 to a low of −58.01 percent in 1937. Note that while the best performance of large and small stocks occurred in the same year, the worst performance did not.

International Stocks: The returns are for MSCI EAFE (Morgan Stanley Capital International for Europe, Australasia, and Far East) for the period 1970–2005. The returns varied from a high of 69.94 percent in 1986 to a low of −23.19 percent in 1990.

Corporate Bonds: These are 20-year loans to high-quality U.S. corporations. The highest return was 43.79 percent in 1982 and the lowest return was −8.09 percent in 1969. Table 1.1 shows that the stock and bond returns do not go hand in hand.

TABLE 1.1   Annual Returns: Stocks, Bonds, Treasury Bills, and Changes in Inflation

Year	Large Company Stocks	Small Company Stocks	International Stocks	Corporate Bonds	Treasury Bills	Inflation
1926	11.62	0.28	—	7.37	3.27	−1.49
1927	37.49	22.10	—	7.44	3.12	−2.08
1928	43.61	39.69	—	2.84	3.24	−0.97
1929	−8.42	−51.36	—	3.27	4.75	0.20
1930	−24.90	−38.15	—	7.98	2.41	−6.03
1931	−43.34	−49.75	—	−1.85	1.07	−9.52
1932	−8.19	−5.39	—	10.82	0.96	−10.30
1933	53.99	142.87	—	10.38	0.30	0.51
1934	−1.44	24.22	—	13.84	0.16	2.03
1935	47.67	40.19	—	9.61	0.17	2.99
1936	33.92	64.80	—	6.74	0.18	1.21
1937	−35.03	−58.01	—	2.75	0.31	3.10
1938	31.12	32.80	—	6.13	−0.02	−2.78
1939	−0.41	0.35	—	3.97	0.02	−0.48
1940	−9.78	−5.16	—	3.39	0.00	0.96
1941	−11.59	−9.00	—	2.73	0.06	9.72
1942	20.34	44.51	—	2.60	0.27	9.29
1943	25.90	88.37	—	2.83	0.35	3.16
1944	19.75	53.72	—	4.73	0.33	2.11
1945	36.44	73.61	—	4.08	0.33	2.25
1946	−8.07	−11.63	—	1.72	0.35	18.16
1947	5.71	0.92	—	−2.34	0.50	9.01
1948	5.50	−2.11	—	4.14	0.81	2.71
1949	18.79	19.75	—	3.31	1.10	−1.80
1950	31.71	38.75	—	2.12	1.20	5.79
1951	24.02	7.80	—	−2.69	1.49	5.87
1952	18.37	3.03	—	3.52	1.66	0.88
1953	−0.99	−6.49	—	3.41	1.82	0.62
1954	52.62	60.58	—	5.39	0.86	−0.50
1955	31.56	20.44	—	0.48	1.57	0.37
1956	6.56	4.28	—	−6.81	2.46	2.86
1957	−10.78	−14.57	—	8.71	3.14	3.02
1958	43.36	64.89	—	−2.22	1.54	1.76
1959	11.95	16.40	—	−0.97	2.95	1.50
1960	0.47	−3.29	—	9.07	2.66	1.48
1961	26.89	32.09	—	4.82	2.13	0.67
1962	−8.73	−11.90	—	7.95	2.73	1.22
1963	22.80	23.57	—	2.19	3.12	1.65
1964	16.48	23.52	—	4.77	3.54	1.19
1965	12.45	41.75	—	−0.46	3.93	1.92
1966	−10.06	−7.01	—	0.20	4.76	3.35
1967	23.98	83.57	—	−4.95	4.21	3.04
1968	11.06	35.97	—	2.57	5.21	4.72

(continued)

TABLE 1.1   (*Continued*)

Year	Large Company Stocks	Small Company Stocks	International Stocks	Corporate Bonds	Treasury Bills	Inflation
1969	−8.50	−25.05	—	−8.09	6.58	6.11
1970	4.01	−17.43	−10.51	18.37	6.53	5.49
1971	14.31	16.50	31.21	11.01	4.39	3.36
1972	18.98	4.43	37.60	7.26	3.84	3.41
1973	−14.66	−30.90	−14.17	1.14	6.93	8.80
1974	−26.47	−19.95	−22.15	−3.06	8.00	12.20
1975	37.20	52.82	37.10	14.64	5.80	7.01
1976	23.84	57.38	3.74	18.65	5.08	4.81
1977	−7.18	25.38	19.42	1.71	5.12	6.77
1978	6.56	23.46	34.30	−0.07	7.18	9.03
1979	18.44	43.46	6.18	−4.18	10.38	13.31
1980	32.42	39.88	24.43	−2.62	11.24	12.40
1981	−4.91	13.88	−1.03	−0.96	14.71	8.94
1982	21.41	28.01	−0.86	43.79	10.54	3.87
1983	22.51	39.67	24.61	4.70	8.80	3.80
1984	6.27	−6.67	7.86	16.39	9.85	3.95
1985	32.16	24.66	56.72	30.09	7.72	3.77
1986	18.47	6.85	69.94	19.85	6.16	1.13
1987	5.23	−9.30	24.93	−0.27	5.47	4.41
1988	16.81	22.87	28.59	10.70	6.35	4.42
1989	31.49	10.18	10.80	16.23	8.37	4.65
1990	−3.17	−21.56	−23.19	6.78	7.81	6.11
1991	30.55	44.63	12.49	19.89	5.60	3.06
1992	7.67	23.35	−11.85	9.39	3.51	2.90
1993	9.99	20.98	32.94	13.19	2.90	2.75
1994	1.31	3.11	8.06	−5.76	3.90	2.67
1995	37.43	34.46	11.55	27.20	5.60	2.54
1996	23.07	17.62	6.36	1.40	5.21	3.32
1997	33.36	22.78	2.06	12.95	5.26	1.70
1998	28.58	−7.31	20.33	10.76	4.86	1.61
1999	21.04	29.79	27.30	−7.45	4.68	2.68
2000	−9.11	−3.59	−13.96	12.87	5.89	3.39
2001	−11.88	22.77	21.21	10.65	3.83	1.55
2002	−22.10	−13.28	−15.66	16.33	1.65	2.38
2003	28.70	60.70	39.17	5.27	1.02	1.88
2004	10.87	18.39	20.70	8.72	1.20	3.26
2005	4.91	5.69	14.02	5.87	2.98	3.42

Treasury Bills: These are short-term loans to the U.S. Treasury. The variability of returns is smaller. The highest return was 14.71 percent in 1981 and the lowest return was virtually zero in the 1938–1940 time frame. The returns have essentially always been positive.

The goal of statistical analysis of capital market history, exemplified by the 80-year period (1926–2005) in Table 1.1, is to uncover the basic relationships in the data to make reasonable predictions regarding the future. By studying the past, one can make inferences about the future. The actual events that have occurred in the past—war and peace, inflation and deflation, oil shocks, market bubbles, and the rise of China and India —may not be exactly repeated in the future, but the event types can be expected to recur. It is sometimes said that the crash of 1929–1932 and the Second World War were the most unusual events. This logic is suspicious, because three of the most unusual events, the market crash of 1987, the high inflation of 1970s and early 1980s, and the market crash of 2008, have occurred in the past three decades. The 80-year history is likely to reveal useful information regarding the future.

1.2 SAMPLE STATISTICS

We consider the data in Table 1.1 to be a sample from a population that includes the past before 1929 and the future after 2005. It is difficult to understand what the data have to convey by simply looking at Table 1.1. We need to summarize the data in a manner conducive to understanding. Three basic summaries are helpful: a plot of the data over time, calculating measures of central tendency (mean and median) and measures of variability (range, variance, standard deviation, and coefficient of variation), and plotting the frequency distribution of the data. These are briefly explained below along with the information each summary provides.

Time Series of Data Figure 1.1(a) shows the time series plot of large company stock returns by year. The returns appear to be randomly fluctuating around a mean. Control chart out-of-control rules, see Chapter 7 and also Joglekar (2003) for an explanation, and the autocorrelation function, described later in this chapter, confirm the conclusion that the yearly returns are randomly distributed. This means that next year's returns cannot be predicted with any deterministic component on the basis of previous years' returns. This is also the case with small company stocks, international stocks, and corporate bonds. This is not a surprising conclusion.

Figure 1.1(b) shows a similar plot for treasury bills. The conclusions here are quite different. When the returns are low, they continue to stay low for some years, and conversely. Control charts and autocorrelation function confirm this conclusion. A substantial portion of next year's returns can be deterministically predicted on the basis of previous years' returns. This is also the case with inflation measured by CPI. This conclusion is useful in making judgments about what the near-term returns and inflation are likely to be.

Statistical Measures of Central Tendency and Variability Let us now turn to the measures of central tendency and variability. Let X denote returns and x_i denote the observed return in year i with x_{max} and x_{min} being the largest and smallest returns. The sample size is denoted by n; in this case, the total number of years $= 80$. Then, the

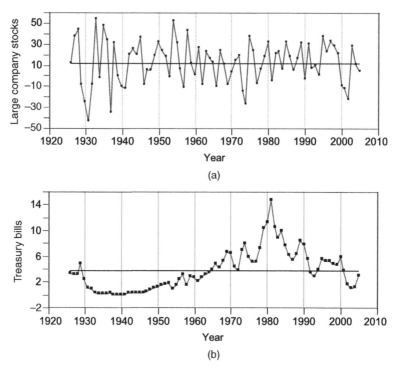

FIGURE 1.1   Time series of (a) large company stock returns by year and (b) treasury bill returns by year.

measures are defined as follows:

$$\text{Average } (\bar{x}) = \frac{\sum x_i}{n}$$

Median = Value of X such that 50% of the observed returns are above
it and 50% of the observed returns are below it

$$\text{Range } (R) = x_{\max} - x_{\min}$$

$$\text{Sample variance } (s^2) = \frac{\sum (x_i - \bar{x})^2}{(n-1)} \tag{1.1}$$

$$\text{Sample standard deviation } (s) = \sqrt{s^2}$$

$$\text{Coefficient of variation } (\text{CV}) = \frac{s}{\bar{x}}$$

Why so many different measures? Each has its role. $\bar{x}$ is the arithmetic average and is usually interpreted as an estimate of the center of the distribution of X. This interpretation is reasonable if the distribution of X is symmetric, such as a normal distribution. If the distribution is asymmetric, mean may be far away from the center and median is a better guide to the central tendency of data.

This difference in mean and median is exemplified by the following example. Some describe the income of the average person in the United States as the average income per person in the United States. This average income has risen by 10 percent between 2000 and 2008. Others describe the income of the average person by lining up incomes from the poorest to the richest and choosing the number in the middle of the lineup. This median income, one may call it the middle-class income, has fallen since 2000. How can the median income decline and the average income go up? The answer is that since 2000 the average income of the bottom 99 percent of the people fell, while the income of the richest 1 percent went up dramatically. The distribution of income is not a symmetric normal distribution and the mean and median are not expected to be equal. If the intent is to find the income such that 50 percent of the people earn less than this income and 50 percent of the people earn more, then median is the right choice.

Range is a simple measure of variability and is the total spread of the data. Range can take only positive values. Large variability results in large range. Range is used frequently, range control charts being one example. However, range depends upon only two observations, the largest and the smallest. If there are outliers, or for some reason a single number is too low or too high (as the 142.87 percent return for small stocks), range will be dramatically affected. We prefer a measure of variability that depends upon all observations.

Variance is one such measure that depends upon all data. It merits some further explanation. It can take only positive values. Variance is defined such that the weight given to an observation is proportional to the square of its distance from the average, and not just linearly proportional to the distance. Variance could have been defined as the absolute average difference $\sum |x_i - \bar{x}|/(n-1)$, where the weights would have been proportional to the distance from the average. Why the quadratic weights? One reason is that variance as defined in Equation (1.1) makes mathematics simpler, as we shall see. The more important reason is that the effects of variability on the customer are disproportionately large as x_i deviates from $\bar{x}$. For a medicinal tablet, if the drug content deviates greatly from the target, the effects of overdose and underdose will be disproportionately large. The denominator in the definition of variance in Equation (1.1) is $(n-1)$. The reason we divide by $(n-1)$ and not n is because, given the value of $\bar{x}$, the numerator in Equation (1.1) has only $(n-1)$ independent terms.

Variance has squared units. For rates of return measured in percent, variance has the units of $(\text{percent})^2$, which are hard to understand. Therefore, standard deviation is defined as the square root of variance with understandable units, same as the original data. The mathematics of variance is simple, but it has the wrong units. The mathematics of standard deviation is complicated, but it has the right units. Therefore, we do all calculations in variance and at the last step take the square root so that the results are in understandable units.

CV is defined as standard deviation divided by the average. At times, CV is the right metric for variability. This is the case when average and standard deviation are

TABLE 1.2   Sample Statistics for Asset Classes and Inflation

Asset Classes and Inflation	Average (%)	Median (%)	Range (%)	Variance (%)^2	Standard Deviation (%)	CV (%)
Large company stocks	12.3	13.4	−43.3 to 54.0	408.1	20.2	164.2
Small company stocks	17.4	19.1	−58 to 142.9	1058.4	32.9	184.7
International stocks	14.4	13.3	−23.2 to 70.0	455.8	21.4	147.7
Corporate bonds	6.2	4.7	−8.1 to 43.8	73.3	8.6	137.3
Treasury bills	3.7	3.2	0.0 to 14.7	9.8	3.1	83.3
Inflation (CPI)	3.1	2.9	−10.3 to 18.2	18.4	4.3	137.1

proportional to each other as often happens when the average varies widely or when we want to compare processes with widely different averages. The inverse of CV is the signal-to-noise ratio, $\bar{x}$ being the signal and s being the noise. Signal-to-noise ratio is used to analyze robustness experiments in Chapter 4.

Table 1.2 summarizes the various measures of central tendency and variability for the various asset classes and inflation in Table 1.1.

What do we learn from Table 1.2? On average, stocks earn more than corporate bonds, which earn more than treasuries. Treasuries barely manage to keep pace with inflation, meaning that if all the money is kept in treasuries, it is not likely to grow by much in real terms over a long time horizon. Small company stocks have the highest average returns. If money could be invested over a long time horizon without having the need to take it out, small company stocks is the place to be. Small company stocks also have the highest risk as measured by the standard deviation (or the range), so the ride will be very bumpy and stomach churning. By looking at the average and standard deviation columns of Table 1.2, we see that the larger the average returns, the larger the standard deviation or risk. This is not unexpected—nothing ventured, nothing gained. In analyzing stocks and bonds, CV is likely to be a better metric because it is nearly constant.

If the median is close to the mean, the underlying distribution of return will be more symmetric. For example, the mean and median for a normal distribution are equal. If we temporarily assume the distribution of returns to be normal, we can interpret the CV as follows. For a CV of 100 percent, the mean and standard deviation are equal. In this case, for a normal distribution of returns, the probability of negative returns in any one year is 16 percent because zero is one standard deviation away from the mean. For a CV of 50 percent, this probability of loss in any one year is about 2.5 percent because zero is two standard deviations away from the mean. Note that none of the asset types have a CV under 50 percent. For a CV of 200 percent, the probability of losing money in any one year is about 30 percent. For small company stocks, while on average the returns are high, there is almost a 30 percent chance of losing money in any one year. By a portfolio we mean investments in stocks, bonds, bills, and so on. How should the portfolio be constructed to give high returns at low risk? As we proceed, we will look at the various aspects of this question, assuming a large time horizon and history as a guide to the future.

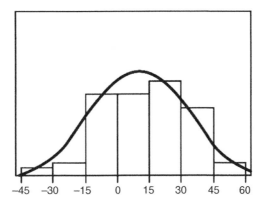

FIGURE 1.2   Distribution of yearly returns of large company stocks (%).

Distribution of Returns   Figure 1.2 shows the frequency distribution of the yearly returns of large company stocks based upon data in Table 1.1. The plot is known as a histogram. The range of returns is divided into cells and the number of observed returns in each cell is plotted on the y-axis. In making judgments with a histogram, it is important to have a large sample size ($n > 100$) and the number of cells k should be determined such that $2^k \approx n$. The distribution appears to be symmetric and bell shaped—a characteristic shape for a normal distribution. Statistical theory usually assumes normality and if the distribution is not normal, actions are taken (data transformation, fitting alternate distributions, etc.) to make theory fit practice. While the measures of central tendency and variability can be calculated regardless of the distribution, if the underlying distribution is non-normal, probability predictions made using a normal distribution will be wrong. Many asset classes have non-normal distribution and we will return to this topic at the end of this chapter.

1.3   POPULATION PARAMETERS

The above discussion focused on sample statistics. Data in Table 1.1 are a sample from the entire population of past and future returns. For the conclusions to be meaningful, a sample must represent the population. A sample is usually a small fraction of the population. The sample average and standard deviation are represented by $\bar{x}$ and s, and their values will change if we take another sample from the population. On the other hand, the entire population has true fixed mean and standard deviation denoted by μ and that is why respectively. These population parameters are usually unknown and unknowable. That is why, we denote the population parameters by Greek alphabets!

Even though the population parameters cannot be exactly determined, decisions are made not on the basis of sample statistics, but on the basis of our understanding of population parameters. For example, one question is: Are the expected yearly treasury bill returns larger than the expected yearly inflation? From Table 1.2, the sample average return for treasury bills is 3.7 percent and it is greater than the observed average inflation of 3.1 percent. However, had we taken a different sample, the results may have been different. So, the answer to our question depends upon whether $\mu_{\text{treasury bills}} > \mu_{\text{inflation}}$ and not upon whether $\bar{x}_{\text{treasury bills}} > \bar{x}_{\text{inflation}}$. The population parameters cannot be known exactly, so the question cannot be answered exactly. However, the population parameters can often be estimated with any desired degree of precision, as we shall soon see. At a practical level, we can usually answer such questions with an acceptably low risk of being wrong.

Normal Distribution   Normal distribution is characterized by two population parameters, mean μ and standard deviation σ, respectively estimated by the sample statistics $\bar{x}$ and s. Figure 1.2 shows the fitted normal distribution for large company stock yearly returns with estimated mean $= 12.3$ percent and estimated standard deviation $= 20.2$ percent as in Table 1.2. Goodness-of-fit tests do not reject the hypothesis of normality in this case. While the histogram in Figure 1.2 shows the frequency distribution of sample data, this fitted normal distribution is our estimate of the underlying population distribution.

If we assume the population mean $\mu = 12.3$ percent and population standard deviation $\sigma = 20.2$ percent, then probability statements can be made based upon the normal distribution. Some probabilities to remember are the following: the interval $\mu \pm 1\sigma$ includes 68 percent of the population, $\mu \pm 2\sigma$ includes 95 percent of the population, and $\mu \pm 3\sigma$ includes 99.7 percent of the population. This means that there is a 95 percent probability that returns in any given year will be in the range of 12.3–40.4 percent. In general, such probabilities can be calculated by first converting the normal distribution to a standard normal with mean zero and standard deviation equal to one and then looking up the probabilities in the normal distribution table in the Appendix. As an example, let X denote the yearly returns of large company stocks. Then, X can be converted to Z, which has the standard normal distribution, by the following transformation:

$$Z = \frac{X - \mu}{\sigma}$$

Suppose we want to find the probability that in any given year large company stocks will have a negative return. Let P denote probability. Then,

$$P(X \le 0) = P\left(Z \le \frac{0-\mu}{\sigma}\right) = P\left(Z \le \frac{0-12.3}{20.2}\right) = P(Z \le -0.61) = 27\%$$

This means that we will expect negative returns from large company stocks roughly 1 in 4 years. Table 1.1 shows that negative returns have occurred with this type of asset 21 times out of 80, close to the predicted probability. A similar calculation done to find the probability that large company stock returns will exceed 5 percent in any given year shows the answer to be 64 percent.

Tolerance Intervals   The above probability calculations are somewhat incorrect because they assume μ and σ to be known. More exact calculations can be done by using the tolerance interval approach, also used in Chapter 5 for setting empirical specifications. We can construct a tolerance interval to include $100(1 - p)$ percent of the population with $100(1 - \alpha)$ percent confidence. The tolerance interval is given by $\bar{x} \pm ks$. The value of k depends upon percent confidence, percentage of the population to be enclosed inside the interval, and the sample size n. The values of k are tabulated in the Appendix. This calculation shows that we are 95 percent sure that the probability of negative returns for large company stocks in any given year is 30 percent, close to the number calculated above assuming μ and σ to be known. The two calculations agree closely because the sample size of 80 is large. For smaller sample sizes, as often happens in the research phase of an industrial project, the probabilities calculated assuming μ and σ to be known and to be unknown can differ widely.

Properties of Variance   It was stated earlier that the mathematics of variance is simpler than the mathematics of standard deviation. The following rules exemplify this simplicity where X and Y denote two types of uncorrelated assets, and a and b denote constants.

$$\text{Variance}(aX) = a^2\text{Variance}(X)$$
$$\text{Variance}(X \pm Y) = \text{Variance}(X) + \text{Variance}(Y) \qquad (1.2)$$
$$\text{Variance}(aX \pm bY) = a^2\text{Variance}(X) + b^2\text{Variance}(Y)$$

The third statement follows from the first two. Note that the variance of a sum or a difference is the sum of variances. Similar equations apply for more than two asset types (or variables) as long as the asset types are uncorrelated with each other. Standard deviation does not have such simple properties.

Reducing Portfolio Risk While Increasing Returns   Consider the case of a conservative investor who would rather invest all the money in corporate bonds ($\mu = 6.2$, $\sigma = 8.6$) because of their lower risk. What will happen if a portion of the funds were invested in the more risky large company stocks ($\mu = 12.3$, $\sigma = 20.2$)? Stocks and bonds are essentially uncorrelated as we shall see later. The results will be as follows:

$$\text{Expected portfolio return} = 6.2a + 12.3(1-a)$$

$$\text{Portfolio variance} = a^2\sigma_X^2 + (1-a)^2\sigma_Y^2$$

$$\text{Portfolio standard deviation} = \sqrt{a^2\sigma_X^2 + (1-a)^2\sigma_Y^2} = \sqrt{74a^2 + 408(1-a)^2}$$

where X denotes the corporate bonds, Y denotes the large company stocks, a is the fraction of portfolio in corporate bonds, and $(1 - a)$ is the fraction of portfolio in large company stocks.

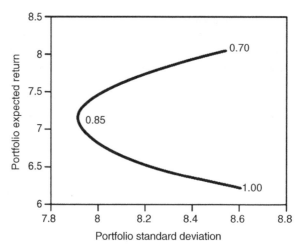

FIGURE 1.3   Expected return and risk for a large company stock and corporate bond portfolio.

The value of a that minimizes the variance can be obtained by differentiating the portfolio variance equation with respect to a and setting the derivative equal to zero. Variance is minimized when

$$a = \frac{\sigma_Y^2}{\sigma_X^2 + \sigma_Y^2}$$

Figure 1.3 shows a plot with the portfolio standard deviation on the x-axis and expected returns on the y-axis for values of a ranging from 1.0 to 0.7. For a 100 percent corporate bond portfolio, $a = 1.0$ and the expected return is 6.2 percent with a standard deviation of 8.6 percent (Table 1.2). As the percentage of large company stock increases from zero, the returns increase and simultaneously the standard deviation reduces. The minimum standard deviation is obtained for $a = 0.85$, meaning a portfolio of 85 percent bonds and 15 percent large company stocks. This portfolio has a return of about 7.1 percent, which is almost a full percentage point higher than that for 100 percent bonds, and a standard deviation of 7.9 percent, which is lower than that for bonds alone. A conservative investor will be better off keeping 15 percent large company stocks in the portfolio. Isn't that interesting!

Distribution of Real Returns   By real returns we mean returns in excess of inflation. If X and Y denote the yearly returns for large company stocks and the yearly inflation, respectively, then

$$\text{Real returns} = X - Y$$

The mean and standard deviation of X are 12.3 and 20.2 percent, respectively. The mean and standard deviation of Y are 3.1 and 4.3 percent, respectively. Using Equation (1.2),

$$\text{Mean (real return)} = 12.3\% - 3.1\% = 9.2\%$$

$$\text{Variance (real return)} = (20.2\%)^2 + (4.3\%)^2 = 426.5(\%)^2$$

$$\text{Standard deviation (real return)} = 20.6\%$$

Note that real returns have a smaller mean and a higher standard deviation compared to the nominal returns denoted by X. The risk of not beating inflation in any given year can be calculated from the properties of normal distribution and is high at 33 percent. The standard deviation of real returns increases because the variance of the difference is the sum of variances.

1.4 CONFIDENCE INTERVALS AND SAMPLE SIZES

The population parameters μ and σ are unknown, but can be estimated within a certain range with a predefined degree of confidence. Let us now turn to the question of calculating such confidence intervals for the population parameters and the question of estimating the associated sample sizes.

Distribution of $\bar{X}$ When referring to the sample average as a variable, we denote it by $\bar{X}$. A specific observed sample average is denoted by $\bar{x}$. The distribution of sample average $\bar{X}$ is an important practical concept in statistics. It is used extensively to compute confidence intervals, conduct hypothesis tests, analyze designed experiments, construct control charts, and so on. It is easy to see that $\bar{X}$, which is the average of n observations, will have a distribution. If we take many samples of size n, the computed $\bar{x}$ values will differ forming a distribution of $\bar{X}$. The following points need to be understood regarding the distribution of $\bar{X}$:

1. As the sample size increases ($n \geq 5$), the distribution of $\bar{X}$ becomes a normal distribution regardless of the distribution of individual values. This statement is also known as the central limit theorem in statistics.
2. The distribution of individual values and the distribution of $\bar{X}$ have the same mean.
3. If the variance of individual values is σ^2, the variance of $\bar{X}$ is σ^2/n. The variance of the average reduces by a factor equal to the sample size that makes up the average. The standard deviation of the average becomes $\sigma/\sqrt{n}$.

These statements can be easily proved using the properties of variance given in Equation (1.2). This exercise is left for the reader. Let us now see how we can use the distribution of the average to obtain a confidence interval for the population mean.

Confidence Interval for μ Even though we cannot know μ precisely, we can bracket it within a certain interval with a prespecified probability. Such intervals are known as confidence intervals for μ. The sample average $\bar{X}$ is normally distributed with mean μ and standard deviation $\sigma/\sqrt{n}$. This means that 95 percent of the observed $\bar{x}$ values are in the interval $\mu \pm 2\sigma/\sqrt{n}$. Thus, we are 95 percent sure that the distance between the observed $\bar{x}$ and μ cannot exceed $2\sigma/\sqrt{n}$ and μ can only be $\pm 2\sigma/\sqrt{n}$ away from $\bar{x}$. Hence,

$$\text{(Sigma known) 95\% confidence interval for } \mu = \bar{x} \pm 2\sigma/\sqrt{n} \qquad (1.3)$$

Equation (1.3) applies when the population standard deviation σ is known. This is rarely the case and σ has to be estimated from the data by calculating the sample standard deviation s. As explained in Chapter 2, this leads to the following practical definition of the confidence interval for μ:

$$\text{(Sigma unknown) 95\% confidence interval for } \mu = \bar{x} \pm t_{0.025, n-1} s/\sqrt{n} \qquad (1.4)$$

where the value of $t_{0.025, n-1}$ is obtained from the t-table in the Appendix corresponding to $(n-1)$ degrees of freedom. Similar intervals can be obtained for any selected level of confidence.

For various categories of assets and for inflation, Table 1.3 shows the 95 percent confidence intervals (CI) for μ. We are 95 percent sure that the true average yearly return for large company stocks is somewhere between 7.8 percent and 16.8 percent. Even though the computed return is 12.3 percent, we cannot say whether the true average return is 8 percent or twice that. This large uncertainty occurs because of the large standard deviation of returns. Confidence intervals can also be used to answer the following questions: Is the yearly mean return for large company stocks smaller than the yearly mean return for small company stocks, and by how much? Does the yearly mean return for treasury bills exceed the expected yearly inflation, and by how much? Chapter 2 discusses such comparative statements in greater detail and shows how the answers can be found by constructing a confidence interval for the difference of means.

Sample Size to Estimate μ Based upon 80 observations, the 95 percent confidence interval for the mean return of large company stocks is 7.8–16.8 percent. From

TABLE 1.3 Confidence Intervals for μ and σ

Asset Classes and Inflation	Average (%)	95% CI for μ (%)	Standard Deviation (%)	95% CI for σ (%)
Large company stocks	12.3	7.8–16.8	20.2	17.6–23.8
Small company stocks	17.4	10.0–24.7	32.9	28.6–38.8
International stocks	14.4	7.2–21.7	21.4	17.3–28.0
Corporate bonds	6.2	4.3–8.1	8.6	7.5–10.1
Treasury bills	3.7	3.1–4.4	3.1	2.7–3.7
Inflation	3.1	2.2–4.1	4.3	3.7–5.1

Equation (1.3), this large uncertainty can be reduced by increasing the sample size n. If $\pm\Delta$ represents the desired width of the 95 percent confidence interval, then from Equation (1.3)

$$\Delta = \frac{2\sigma}{\sqrt{n}} \quad \text{and} \quad n = \frac{4}{d^2} \quad \text{where } d = \frac{\Delta}{\sigma} \tag{1.5}$$

If we want to estimate the true yearly mean return for large company stocks within 2 percent of the true value, then $\Delta = 2$ percent. For $\sigma = 20.2$ percent, $d \approx 0.1$ and $n = 400$. We will need 400 years of data to estimate the mean yearly return of large company stocks within 2 percent of the truth! While this cannot be accomplished here, in industrial problems, additional data can always be obtained if required. Equation (1.5) shows how sample size can be determined *before collecting the data* once the values of Δ and σ are specified. Δ is based upon business considerations, namely, how well do we need to know μ to make the decision at hand, and σ is estimated from prior information. Note that the sample size depends upon the ratio of Δ and σ and not on their individual values.

Confidence Interval and Sample Size for σ For population standard deviation, the exact computation of confidence interval and sample size is more complicated (see Joglekar (2003)). Table 1.4 summarizes the results assuming the data to be normally distributed. It may be used as follows. The computed standard deviation s for large company stocks is 20.2 percent based upon a sample size of 80. For this sample size, the 95 percent confidence limit for σ goes from 87 to 118 percent of s (Table 1.4). We are 95 percent sure that the true value of σ is in the interval 17.6–23.8 percent. Note that the confidence interval for σ is asymmetrical around the observed sample standard

TABLE 1.4 Sample Size and 95% Confidence Limits for $100(\sigma/s)$

Sample Size	Lower Limit	Upper Limit
3	52	626
4	57	373
6	62	245
8	66	204
10	69	183
16	74	155
21	77	144
31	80	134
41	82	128
51	84	124
71	86	120
101	88	116
200	90	110
500	94	106
1000	96	104
5000	98	102

deviation because the distribution of s is not symmetric. The 95 percent confidence intervals for the population standard deviation for the various asset classes and inflation are shown in Table 1.3.

As can be seen from Table 1.4, the confidence interval for σ is very wide for small sample sizes. The table can also be used to decide the sample size necessary to estimate σ with the desired precision. Approximately 200 observations are necessary to estimate σ within ± 10 percent of the true value.

1.5 CORRELATION

The properties of variance, described by Equation (1.2) for uncorrelated variables and used to demonstrate how risk reduction occurs by combining various asset classes, change if the variables (or asset classes) are correlated with each other.

Correlation Coefficients   Figure 1.4 shows a scatter diagram of large and small company stock returns. Large returns for large and small company stocks seem to go somewhat hand in hand and vice versa. Correlation coefficient is a measure of the strength of the linear relationship between two asset classes or generally between two variables. The calculated value of the correlation coefficient is denoted by R and the true value by ρ.

The correlation coefficient R can take values between -1 and $+1$. If all the plotted points are on a perfect straight line with a positive slope, $R = 1$. If all the points are on a perfect straight line with a negative slope, $R = -1$. If the plot looks like a circle, it would indicate no correlation and $R = 0$. Specifically, correlation coefficient between X and Y is given by

$$R = \frac{s_{xy}}{s_x s_y}, \quad \text{where } s_{xy} = \frac{1}{n-1} \sum (x_i - \bar{x})(y_i - \bar{y}) \tag{1.6}$$

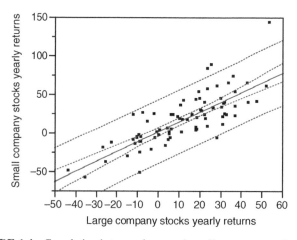

FIGURE 1.4   Correlation between large and small company stock returns.

TABLE 1.5 Correlation Structure for Asset Classes and Inflation

Asset Classes and Inflation	Large Stocks	Small Stocks	International Stocks	Corporate Bonds	Treasury Bills	Inflation
Large stocks	1.0					
Small stocks	0.8	1.0				
International stocks	0.5	0.4	1.0			
Corporate bonds	0.2	0.1	0.1	1.0		
Treasury bills	0.0	−0.1	−0.2	0.2	1.0	
Inflation	0.0	0.0	−0.3	−0.2	0.4	1.0

where s_x and s_y are the computed standard deviations of X and Y. s_{xy} is called the covariance between X and Y and measures the tendency of X and Y to go hand in hand. The correlation coefficient is the covariance standardized by the standard deviations. The square of the correlation coefficient measures the fraction of the variance of Y that is explained by changes in X. If $|R| = 1.0$, all the variability in Y will be explained by the variability in X. This leads to the definition of the values of correlation coefficient that are practically important. $|R| < 0.5$ is considered weak, $|R| > 0.8$ is considered strong, and $0.5 < |R| < 0.8$ is considered moderate correlation. A correlation coefficient is statistically significant, that is, $\rho \neq 0$, if $t = R\sqrt{n-2}/\sqrt{1-R^2}$ exceeds the critical value for t_{n-2}.

Table 1.5 summarizes the correlation coefficients for all asset classes and inflation. Approximately, $|R| > 0.25$ is statistically significant in this table, which is based upon a sample size of 80. There is a strong positive correlation between small and large company stocks. International stocks are moderately positively correlated with large and small company stocks. Treasury bills are weakly correlated with inflation. Most other correlations are statistically and practically small. There are no strong negative correlations.

Regression Analysis Just because the yearly return for the large and small company stocks are correlated does not mean that there is necessarily a causal relationship between them. Correlation does not prove causation, although many a time, there is an underlying causal relationship. In the case of large and small company stocks, the likely scenario is that both go up or down in response to the perceived economic conditions and other factors. Such factors that induce a correlation are called lurking variables. Regression analysis allows us to build an equation describing the relationship, whether causal or otherwise (see Chapter 10 for further details). The equation is

Small company stocks yearly returns $= 1.59 + 1.28$ Large company stocks

yearly returns $+$ Residual

Figure 1.4 shows a plot of this linear relationship. Residual is the portion of small company stocks yearly returns that is not explained by the corresponding large company stocks yearly returns. The intercept and slope in the above equation are determined to minimize the residual sum of squares and this approach to estimate the coefficients of the equation (also called the model) is known as the linear least squares approach. In Figure 1.4, the inner hour glass looking lines represent the 95 percent confidence interval on the fitted line, meaning that we are 95 percent sure that the true relationship is inside these lines. The outer two lines indicate the region within which 95 percent of the individual points are expected to lie.

Building such equations is an important part of the application of statistical methods to practical problems. We will encounter the subject repeatedly in several future chapters. In particular, the subject of model building is considered in greater detail in Chapter 10.

Autocorrelation   The correlation coefficients considered before are between two different variables or assets. Correlations can occur within a single asset, for example, between the return at time t and return at time $(t-1)$, $(t-2)$, and so on, and are called autocorrelations of lag 1, lag 2, and so on. Many statistical analyses assume that successive observations are uncorrelated, that is, all autocorrelations are zero. For example, the out-of-control rules for control charts are based upon such an assumption. It is therefore a good practice to check the assumption by computing the autocorrelation function.

Let us suppose that we want to compute the autocorrelation function for treasury bill returns. Equation (1.6) can be used to compute the autocorrelation k lags apart by defining x to be the return at time t and y to be the return at time $(t-k)$. Autocorrelation function is simply the plot of these autocorrelations for all values of k.

It turns out that all asset classes have insignificant autocorrelations except treasury bills. It was shown earlier in Figure 1.1(b) that the time series of treasury bill returns has a slow cyclic behavior. The computed autocorrelation coefficient between return in year t and return in year $(t-1)$ is approximately 0.92 and is statistically and practically very significant. The computed autocorrelations k lags apart are approximately $(0.92)^k$. The presence of these autocorrelations means that the return at time t can be predicted to some degree based upon the known previous returns. Equations describing such relationships are known as time series models. For the treasury bill return data, the observed pattern of autocorrelations suggests what is known as a first-order autoregressive model shown below, which allows the prediction of next year's return based upon the previous year's return.

$$\text{Treasury return } (t) = 0.92 \text{ Treasury return } (t-1) + \text{Residual}$$

1.6   PORTFOLIO OPTIMIZATION

An important application of the statistical analysis of returns is the mean–variance optimization of a portfolio. The following assumptions are made:

1. The expected return is what investors find desirable.
2. Risk is measured by the expected standard deviation of returns and is what investors want to reduce.

3. The relationship between different assets is characterized by the correlation coefficients between every pair of assets.

Properties of Variance   The properties of variance described by Equation (1.2) apply for uncorrelated asset classes. When there are correlations, the equations change. Let X and Y represent two asset classes, and let a be the fraction of the portfolio invested in asset class X. Then,

$$\text{Expected portfolio return} = a\mu_X + (1-a)\mu_Y$$

$$\sigma^2_{\text{portfolio}} = a^2\sigma^2_X + (1-a)^2\sigma^2_Y + 2a(1-a)\sigma_X\sigma_Y\rho_{XY} \tag{1.7}$$

Equation (1.7) can now be applied to mixtures of asset classes to determine the expected returns and risks of any portfolio. Let us consider some specific situations.

1. When X and Y are perfectly positively correlated, that is, $\rho_{XY} = 1.0$, then

$$\sigma_{\text{portfolio}} = a\sigma_X + (1-a)\sigma_Y$$

   The portfolio standard deviation is a weighted sum of the two standard deviations.
2. When X and Y are uncorrelated, that is, $\rho_{XY} = 0$, then

$$\sigma_{\text{portfolio}} = \sqrt{a^2\sigma^2_X + (1-a)^2\sigma^2_Y}$$

   The standard deviation is minimized when

$$a = \frac{\sigma^2_Y}{\sigma^2_X + \sigma^2_Y}$$

   This case applies to all the asset classes that are nearly uncorrelated in Table 1.5.
3. When X and Y are perfectly negatively correlated, that is, $\rho_{XY} = -1.0$, then

$$\sigma_{\text{portfolio}} = a\sigma_X - (1-a)\sigma_Y$$

   Note that $\sigma_{\text{portfolio}}$ can never be negative. However, in a generally unrealistic case, it can become zero when

$$a = \frac{\sigma_Y}{\sigma_X + \sigma_Y}$$

   Reducing the risk to zero while obtaining high returns will be wonderful. Unfortunately, strongly negatively correlated assets with large expected returns are difficult to find.

Let us now consider what the results will be if the portfolio consists of equal amounts of large and small company stocks. In this case, $a = 0.5$.

$$\text{Expected portfolio return} = 0.5\,(12.3\%) + 0.5\,(17.4\%) = 14.85\%$$

Substituting $\sigma_X = 20.2\%$, $\sigma_Y = 32.9\%$, and $\rho_{XY} = 0.8$ in Equation (1.7),

$$\sigma^2_{\text{portfolio}} = 638(\%)^2 \quad \text{and} \quad \sigma_{\text{portfolio}} = 25.3(\%)$$

Equation (1.7) generalizes to a mixture of any number of asset classes and allows us to examine the risk and return consequences of the portfolio as illustrated below.

Mean–Variance Portfolio Optimization   Let us suppose that our objective is to get a 12 percent expected return with the smallest possible variance. We can earn a little better than 12 percent expected return if we invest the entire portfolio in large company stocks. However, the risk as measured by the standard deviation will be over 20 percent. Can we do better in terms of risk by diversifying the portfolio?

Let us say that we consider investing in three asset classes: large company stocks, international stocks, and corporate bonds. We want to know if a portfolio of these asset classes can be constructed to give us the desired returns at a lower risk.

Table 1.6 shows the mean, standard deviation, and percent CV of return for seven portfolios. Three portfolios consist of individual assets only, three have an equal mixture of two asset classes, and one has an equal mixture of all three asset classes. The expected returns were computed using Equation (1.7) and values from Tables 1.2 and 1.5.

Each of the seven portfolios is a mixture in the sense that the three asset classes always add to 1 (or 100%). As shown in Figure 1.5, such a three-component mixture can be represented by a triangle, which has the property that at any point inside the triangle the sum of the three components is always 1. Therefore, the infinite possible portfolios are contained inside the triangle. Our objective is to find the portfolio that has a mean yearly return of 12 percent with the smallest standard deviation.

In such cases, contours of constant response can be generated by first building an equation, for example, for the mean return as a function of the three asset classes based upon data in Table 1.6. The methodology is explained in Chapter 3 on design of experiments. Presently, let us simply consider what the results in Figure 1.5(a)–(c) suggest.

Figure 1.5(a) shows the response surface contours for mean return. The corner points of the triangle are individual assets. All the seven portfolios are shown by dots. All portfolios that fall on the contour labeled 12 have a 12 percent mean yearly return.

TABLE 1.6   Returns and Risks for Seven Portfolios

Portfolio	Large Company Stocks	International Stocks	Corporate Bonds	Mean (%)	Standard Deviation (%)	CV (%)
1	1.0	0.0	0.0	12.3	20.2	164.2
2	0.0	1.0	0.0	14.4	21.4	147.7
3	0.0	0.0	1.0	6.2	8.6	137.3
4	0.5	0.5	0.0	13.35	18.0	134.8
5	0.5	0.0	0.5	9.25	11.7	126.5
6	0.0	0.5	0.5	10.3	11.9	115.5
7	0.33	0.33	0.33	11.0	12.8	116.4

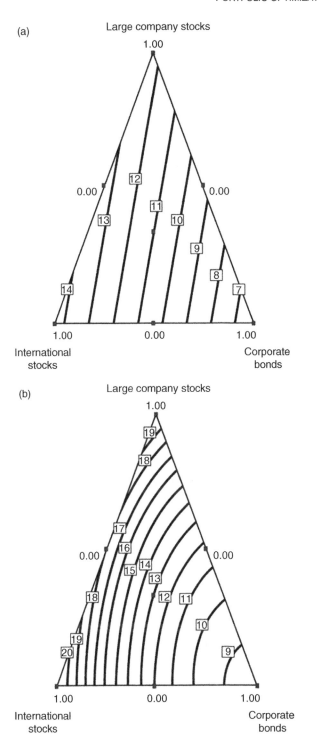

FIGURE 1.5   Response surface contours for (a) portfolio expected return, (b) portfolio standard deviation, and (c) portfolio CV. (d) Overlay for portfolio expected return and standard deviation.

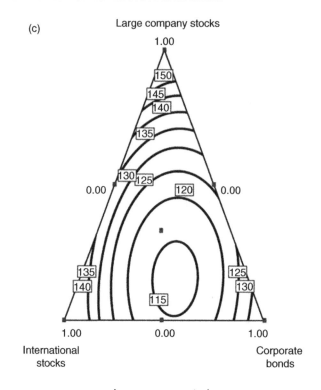

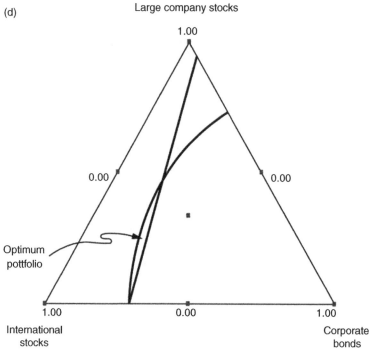

FIGURE 1.5 (*Continued*).

The contours are straight lines (because the equation for mean return is a simple linear equation with no cross-product terms) indicating a rising plane from right to left. Figure 1.5(b) shows the contours for standard deviation. The contours are curved because of the cross-product terms, again rising from right to left. Figure 1.5(c) shows the contours for CV. The contours look like a bowl with minimum CV in the lower middle portion of the triangle.

If the 12 percent mean return line from Figure 1.5(a) is drawn on the standard deviation contours in Figure 1.5(b), it is seen that the portfolios on the 12 percent mean return line have different standard deviations, ranging from 14.5 to 19 percent. The portfolio with minimum standard deviation can be obtained by simply superimposing Figure 1.5(a) and (b) and finding the point on the 12 percent mean return line where the standard deviation is a minimum. Figure 1.5(d) shows such an overlay plot and the portfolio consisting of 27 percent large company stocks, 51 percent international stocks, and 22 percent bonds is the optimum portfolio with expected returns exceeding 12 percent and with a standard deviation less than 14.6 percent. This is a substantial risk reduction when compared with the 20.2 percent standard deviation for large company stocks alone.

Forecasting Wealth   The discussion so far has focused on the yearly returns of a portfolio. How can we forecast the returns several years out? This cannot be done by simply using the expected yearly rate of return (μ) and saying that the expected wealth n years out will be $(1 + \mu)^n$ times the initial investment. This may be understood as follows. If the first year return is 100 percent and the second year return is -50 percent, the average yearly return over two years will be computed as 25 percent. However, in reality, after two years one would have the same amount of money as one started with. So, the _annualized_ return is zero percent. The average return approach does not work for multiple years. The second reason is that the investors are, or should be, interested not only in a point forecast of wealth in the future, but also in the probability distribution of wealth in the future. The following approach may be used to make predictions (see Ibbotson (2006)).

If W_0 is the initial investment, W_n is the wealth n years out, and x_i is the rate of return in year i, then

$$W_n = W_0(1+x_1)(1+x_2) \cdots (1+x_n)$$

Taking the natural logarithm represented by ln,

$$\ln W_n = \ln W_0 + \ln(1+x_1) + \ln(1+x_2) + \cdots + \ln(1+x_n)$$

If $(1 + x_i)$ has a log-normal distribution, then $\ln(1 + x_i)$ will be normally distributed. Even if $(1 + x_i)$ does not have a lognormal distribution, for large time horizon ($n > 5$), the central limit theorem says that the distribution of $\ln W_n$ will be close to normal, meaning that W_n will have a log-normal distribution.

Let x_g represent the compound growth rate over the same time period. Then,

$$x_g = \left(\frac{W_n}{W_0}\right)^{1/n} - 1$$

If W_n has a log-normal distribution, then it follows that $(1 + x_g)$ will have a log-normal distribution.

To use this log-normal model, given the portfolio expected return μ and standard deviation σ, we must first calculate the mean M and standard deviation S of $\ln(1 + x_i)$ as follows:

$$M = \ln(1+\mu) - \frac{S^2}{2}$$

$$S = \sqrt{\ln\left[1 + \left(\frac{\sigma}{1+\mu}\right)^2\right]} \tag{1.8}$$

Equation (1.8) results from the properties of the log-normal distribution. It then follows that $\ln W_n$ has a normal distribution with mean $\ln W_0 + Mn$ and standard deviation $S\sqrt{n}$. To calculate a particular percentile of the wealth distribution for a given time horizon, we need the Z-score corresponding to that percentile. For example, the Z-score corresponding to the 95th percentile is 1.645, corresponding to the 50th percentile is zero, and corresponding to the 5th percentile is -1.645. This can be seen from the normal distribution table in the Appendix. From the distribution of $\ln W_n$, it follows that

$$W_n = W_0 \, e^{Mn + ZS\sqrt{n}}$$

$$x_g = e^{M + Z(S/\sqrt{n})} - 1 \tag{1.9}$$

For the optimal portfolio derived in the previous section, $\mu = 0.12$ and $\sigma = 0.146$. Substituting these in Equation (1.8) gives $M = 0.105$ and $S = 0.13$. If $10,000 is invested in this portfolio for a period of 50 years, then $W_0 = 10,000$ and $n = 50$. Substituting these in Equation (1.9), we find that there is a better than 95 percent chance that the initial $10,000 will become more than $420 thousand, a 50 percent chance that the wealth will exceed $1.9 million, and a 5 percent chance that the wealth will exceed $8.64 million in then year dollars. It is a large amount even in today's dollars, a computation that is left for the reader. It has been said that everybody's grandchildren ought to be rich. Unfortunately, most of us either do not, or do not have the luxury to, make the necessary investments for our potential grandchildren 50 years before they go to college!

1.7 QUESTIONS TO ASK

The following is a comprehensive list of questions on topics covered in this chapter. While answers for many of these questions are provided in this chapter, some of the questions require reading the next chapters and also references such as Joglekar (2003).

1.7.1 What Data to Collect?

1. What is the objective of data collection? Is it focused? Are there conflicting objectives?
2. Which factors are included in the study? Which factors are outputs, which are inputs, and which are noise? Are the factors continuous or attribute?
3. Should other output, input, or noise factors be included?
4. Can attribute measurements be converted to continuous measurements?
5. How will the study be planned? Are design of experiment principles being used? Are the sample sizes properly determined? Have the details of conducting the study been well thought through?
6. What methods will be used for analysis? Are the decision-making criteria clearly defined? Will the analysis answer the objectives of data collection? Are the assumptions made by the various contemplated analyses likely to be satisfied?

1.7.2 Sample Statistics

1. Is the distribution expected to be normal? If not, why not?
2. What are the results of the normality test? Are there outliers? If the distribution is not normal, what is the likely impact on the analysis, and are actions such as data transformation necessary before data analysis?
3. Is it more appropriate to use the mean or the median? Standard deviation or CV?

1.7.3 Population Parameters (Confidence Intervals and Sample Sizes)

1. How well does the mean or the standard deviation need to be estimated for decision-making purposes?
2. Is the sample size properly determined to estimate mean and standard deviation with the desired precision? Is the precision with which the mean and standard deviation should be estimated well defined? Is a prior estimate of variance available to help estimate the sample size? Are the two risks of wrong decisions well chosen? What can be done to reduce the need for a large sample size? Are ideas such as accelerated testing, worst-case testing, and designed experiments relevant to reduce the sample size?
3. If the study was conducted with arbitrarily selected sample size, or with happenstance data, what are the adverse consequences?
4. For continuous data, are the assumptions of normality, independence, and constant variance satisfied?
5. For discrete data, is the distribution as expected, for example, binomial or Poisson?
6. Are the confidence intervals being correctly interpreted in making decisions?

1.7.4   Tolerance Intervals

1. What is the specific purpose for computing the tolerance interval, for example, is the purpose to compute fraction defective or to set empirical specifications?
2. What is the basis for the percent confidence and percentage of the population used to compute the value of k?
3. How was the sample size selected?

1.7.5   Correlation and Regression

1. Is the purpose of conducting the correlation and regression analysis clearly stated?
2. For happenstance data, have the pitfalls of such an analysis (correlation without causation, inadequate ranges of input factors, stratification, and parabolic relationships) properly considered?
3. Are the statistical and practical significance of correlation coefficients correctly interpreted?
4. Has the model been properly fitted and diagnostically checked? Does the model explain a large proportion of the variability of output and is the prediction error sufficiently small? Is the unexplained variability of output due to measurement variability or due to factors not included in the model? Does the residual analysis show lack of fit? Have data transformations been considered where necessary?
5. If the data have been collected in a time series fashion, is there a large autocorrelation? If so, what is the likely impact on the contemplated analysis such as control charts? What corrective measures are necessary?

Further Reading   Ibbotson (2006) provides historical data on stocks, bonds, treasury bills, and inflation. It also provides useful practical information on asset classes, risk reduction, and wealth projection. It is a good reference to pursue the subject of portfolio management. The book by Joglekar (2003) provides detailed information on various statistical topics covered in this chapter including distributions, confidence intervals, sample sizes, tolerance interval, correlation, and regression. The material is presented with more conventional industry examples.

CHAPTER 2

WHY NOT TO DO THE USUAL t-TEST AND WHAT TO REPLACE IT WITH?

It was almost 100 years ago that the t-distribution and the t-test were invented by W. S. Gosset. This important development provided the statistical basis to analyze small sample data. One application of the t-test in industry today is to test the hypothesis that two population means are equal. Decisions are often made purely based upon whether the difference is signaled as statistically significant by the t-test or not. This application of the t-test to industrial practical problems has two bad consequences: practically unimportant differences in mean may be identified as statistically significant and potentially important differences may be identified as statistically insignificant. This chapter shows that these difficulties cannot be completely overcome by conducting another type of t-test, by computing sample sizes, or by conducting postexperiment power computations. For practical decision-making, replacing the t-test by a confidence interval for difference of means resolves these difficulties. Similar arguments apply to paired t-test, the F-test, and all other common hypothesis tests.

Many of you have done a t-test. Some of you use it routinely to demonstrate that a change in process mean has occurred, that two test procedures are equivalent, that one supplier may be replaced by another, that a factor is producing an effect, and that one ingredient may be substituted for another, and also in many other applications. You may have listened to your coworkers state that "the t-test shows a statistically significant difference between the old product design and the new product design." The standard operating procedures (SOPs) in your company may mandate the use of a t-test to demonstrate equivalence. With today's software packages, a t-test is only a click of the mouse away. It is a very popular test!

Industrial Statistics: Practical Methods and Guidance for Improved Performance   By Anand M. Joglekar
Copyright © 2010 John Wiley & Sons, Inc.

A Tale of Two Companies   Some years ago, I received a call from a project manager who was involved with the production of patches that transfer drug through skin. His team had the objective of increasing the productivity of the manufacturing line. Over the previous several months, this objective had been accomplished through many individual improvements and capital investments in new machines. Production rate had been doubled which meant large operating-cost reductions. While all of this was good, their *t*-test to demonstrate that the product produced by the new production line was equivalent to the old product had failed! As a result, considerable work was currently ongoing and anticipated, requiring more than a 3-month delay and large expenditures. The project manager did not dispute that the *t*-test, mandated by the standard operating procedure, had failed, but felt that the difference in the mean drug-transfer rates was too small to be of practical consequence and should be ignored.

This company had a standard operating procedure which stated that any changes to the manufacturing process must be verified by conducting a *t*-test to demonstrate the product itself had not changed. In this particular case, a large amount of *in vitro* drug-transfer data was available for products produced by the old process, and similar large amounts of data had been collected on products produced by the new process. When these two data sets were used to conduct a *t*-test for difference of means, the test had failed, meaning that there was a statistically significant difference between the mean drug transfer for products produced by the old process and the new process. This conclusion had brought the commercialization of the new process to a halt.

Another case with a different company involved the transfer of a measurement system from R&D to manufacturing. When such a method transfer takes place (see Joglekar (2003)), it is usual to conduct a statistical test to demonstrate that people in manufacturing are well trained in the use of the measurement system and are getting the same results as were obtained in R&D. One common approach to demonstrate this equivalence is to ask R&D and manufacturing to each measure (say) the same 10 products or 2 separate sets of 10 randomly selected products from the production line and then use a *t*-test to show that the difference between the population means in R&D and manufacturing is statistically insignificant. In this particular instance, this had indeed been the case and the method transfer was deemed to be a success. Everybody connected with the method transfer project was happy.

As more data accumulated in manufacturing, it appeared that the average value of the measured dimension had shifted compared with that in R&D. Initially, it was thought that the manufacturing process had caused this shift, but eventually suspicion grew that perhaps the measurement system was at fault. A second, larger method transfer study was conducted to compare R&D measurements with manufacturing measurements of the same products. It turned out that the manufacturing personnel were aligning the product differently from what was done in R&D. This difference had caused a shift in the mean measurement.

In the two cases cited above, two different wrong conclusions were reached, each resulting in a large amount of nonvalue-added work and financial losses. In the case of patches, a small practically unimportant difference was identified by the *t*-test as statistically significant and the standard operating procedure forced additional work when none was needed. In the measurement method transfer study, the initial *t*-test showed the difference between R&D and manufacturing results to be statistically insignificant, when the difference was practically important.

What were the reasons for these wrong conclusions? Were these wrong applications of the *t*-test? Was the wrong *t*-test conducted? Was the collected data too little or too much for a satisfactory analysis? Perhaps, the standard operating procedure was too stringent in bringing the project to a halt. Or, do these examples illustrate the inherent limitations of the *t*-test as it is commonly practiced in industry?

2.1 WHAT IS A *t*-TEST AND WHAT IS WRONG WITH IT?

A *t*-test is often used in industry to assess if two population means are the same or are different. This is routinely done by conducting the test on available or purposefully collected data. If the *t*-test shows that the difference in population means is statistically significant, then the two means are taken to be different, otherwise not.

Let us consider an example. Suppose that we have a current package (design A) with satisfactory seal strength. A new package (design B) is available and is much cheaper. We would prefer to switch to this new package because the cost savings will be considerable. However, to use the new package, it needs to have the same seal strength as the old design. If the seal strength is much smaller, the package may inadvertently open too easily. If the seal strength is much larger, the package may be difficult to open. Thus, if the seal strength for the new design is smaller or larger than the seal strength of the old design, there will be customer dissatisfaction with the new design and the resultant financial losses will exceed the projected cost savings. For our purposes here let us only consider the question of demonstrating that the mean seal strengths of the two packages are equivalent.

Let us suppose that we measure the seal strengths of 10 randomly selected packages made with the old design A and 10 randomly selected packages made with the new design B and obtain the data reported in Table 2.1.

Note that the average, standard deviation, and variance in Table 2.1 are sample estimates and, respectively, estimate the *population* parameters μ, σ, and σ^2. In theory, we could produce an infinite number of packages with design A and measure their seal strengths. The average and standard deviation of these infinite measurements are the population mean μ_A and population standard deviation σ_A for design A and population mean μ_B and population standard deviation σ_B for design B. Since we cannot actually produce and measure these infinite products, the true population means and standard deviations are unknown.

The two designs will be exactly identical in terms of their seal strength if their population means and standard deviations are identical. A *t*-test, as usually practiced, is a test of the hypothesis that the two population means are equal, against the alternate hypothesis that the population means are not equal. Based upon the data, *t*-test shows whether we can be sufficiently confident in saying that the two unknown population means are different.

TABLE 2.1 Seal Strength Data

	Design A	Design B
	2.79	3.11
	2.38	3.37
	2.49	3.74
	1.92	3.48
	3.31	2.68
	2.55	3.12
	2.22	3.79
	2.51	2.31
	3.32	2.29
	2.78	2.66
Average	$\bar{x}_A = 2.627$	$\bar{x}_B = 3.055$
Standard deviation	$s_A = 0.422$	$s_B = 0.551$
Variance	$s_A^2 = 0.190$	$s_B^2 = 0.303$

The problem of determining whether the two population means are different or not is made more complicated by the fact that the true population standard deviations are also unknown. This problem was solved by W. S. Gosset almost a hundred years ago. He was a chemist working for Guinness brewery, and was often faced with the analysis of small sample data. The sample sizes were necessarily small because of the expense of conducting large-scale brewing experiments. The method then in use was to treat sample standard deviation as if it were the population standard deviation (reasonable for very large sample sizes, not reasonable for small sample sizes). Gosset solved the small sample size problem by inventing the *t*-distribution and the *t*-test. Gosset published the results under the pseudonym Student, hence, the terms Student's *t*-distribution, Student's *t*-test, and studentization.

Without loss of generality, for purposes of current discussion, let us assume that the two population variances are equal and that the main question is whether the two population means are equal or not. We can then obtain a better estimate of population variance (0.247) by averaging the two sample variances in Table 2.1. A pooled estimate of population standard deviation σ is obtained by taking the square root of 0.247, that is, $s = 0.5$.

It seems clear that if the difference between the two population means is large, then the difference between the two sample averages is likely to be large and vice versa. How to judge whether the observed difference between the two sample averages is large or small? What yardstick should we use? For simplicity, we first assume σ to be known. Then, we may proceed as follows. If the two population means are equal, namely, $(\mu_A - \mu_B) = 0$, then the observed difference between the sample averages $(\bar{x}_A - \bar{x}_B)$ will have a normal distribution with mean $= 0$ and standard deviation $= \sqrt{2}\sigma/\sqrt{n}$, where n is the number of observations for each design (in our case $n = 10$). This can be seen from the properties of variance explained in Chapter 1, namely, by knowing that the variance of an average is $1/n$th the variance of individual values and that the variance of a difference is the sum of variances. This means that

$$Z = \frac{\bar{x}_A - \bar{x}_B}{\sqrt{2}\sigma/\sqrt{n}} \tag{2.1}$$

will have the standard normal distribution with mean $= 0$ and standard deviation $= 1$. From Equation (2.1), given the sample averages and the sample size, the value of Z can be calculated if σ is assumed to be known. It is well known that the probability that Z is within ± 2 is 95 percent. Therefore, if our assumption $(\mu_A - \mu_B) = 0$ is correct and if the calculated value of Z is outside the interval ± 2, then we will be more than 95 percent sure that the two means μ_A and μ_B are not equal. Thus, the Z-test given in Equation (2.1) provides us with a yardstick to judge whether the two means are statistically equal or not.

The assumption that the population standard deviation is known may be reasonable for large values of n because then the calculated standard deviation will be close to the population standard deviation. For small n, Gosset replaced the unknown σ by the calculated s and showed that the calculated

$$t = \frac{\bar{x}_A - \bar{x}_B}{\sqrt{2}s/\sqrt{n}} \tag{2.2}$$

has a t-distribution with $2(n - 1)$ degrees of freedom. The t-distribution is a symmetric, bell-shaped distribution, but has fatter tails compared with the normal distribution. Therefore, the interval that captures 95 percent probability becomes larger than ± 2. The necessary values are tabulated in the t-distribution table in the Appendix. As the degrees of freedom increase, that is, as we gather larger amounts of data, the t-distribution becomes the Z-distribution.

Presently, the degrees of freedom are 18 and the interval ± 2.1 (obtained from the table of t-distribution in the Appendix) captures 95 percent of the probability. Note that this is only slightly bigger than ± 2, which we would have otherwise used. This means that the t-distribution converges to the Z-distribution rapidly and it is only for very small sample sizes that the difference between a t-distribution and a Z-distribution is pronounced. From Equation (2.2) and Table 2.1, the calculated value of t is $- 1.92$. Since this value is not outside the ± 2.1 interval, we cannot say with 95 percent confidence that the two designs have different mean seal strengths. For all practical purposes, we accept the hypothesis that the two designs have the same mean seal strength. The application of the t-test would lead us to switch to the cheaper design B.

What Is Wrong with the t-Test?   Note that the way a t-test is usually structured, it answers the question: Based upon the data, can we be sufficiently confident in saying that the two population means are different? It tests the hypothesis $(\mu_A - \mu_B) = 0$ against all alternatives. However, we do not need data to tell us that the two designs have different population means. No two things are exactly alike. We know that the two population means differ, if in the 100th decimal place! So the usual t-test answers the wrong question! "Are two population means exactly equal?" is hardly a question of practical interest in industry.

What we really want to know is not whether the two population means are different, we know the answer to that question, but rather, *how different are they?* Presently, our real question is: How big is the difference between the true mean seal strengths of the two designs? We cannot answer this question without data. Depending upon the answer, and our definition of practically important difference, we can then decide which package design to use. A *t*-test as usually conducted does not provide the answer that we seek and, therefore, should not be used.

2.2   CONFIDENCE INTERVAL IS BETTER THAN A *t*-TEST

A *t*-test should be replaced by confidence interval for $(\mu_A - \mu_B)$. The confidence interval answers the question of interest, namely, how big is the difference between the two population means?

The 95 percent confidence interval for $(\mu_A - \mu_B)$ is given by

$$(\bar{x}_A - \bar{x}_B) \pm t_{2(n-1),0.025}\left(\sqrt{2}s/\sqrt{n}\right) \tag{2.3}$$

As you can see, this merely represents a rearrangement of the *t*-test in Equation (2.2). If you can do a *t*-test, you can calculate the confidence interval and vice versa. In this sense, they are equivalent. Presently, substituting in Equation (2.3) from Table 2.1, the calculated 95 percent confidence interval is $(-0.90$ to $+0.04)$. This confidence interval is interpreted to mean that we are 95 percent sure that the true difference $(\mu_A - \mu_B)$ is in the range $(-0.90$ to $+0.04)$.

Since zero is inside this confidence interval, the assertion that the two means are equal cannot be disputed based upon the data. Thus, conducting a *t*-test for equality of means is the same as asking whether zero is inside the confidence interval for $(\mu_A - \mu_B)$. If the confidence interval includes zero, the difference is statistically insignificant, otherwise, the difference is statistically significant. The confidence interval and the *t*-test come to the same conclusion: We cannot say with 95 percent confidence that the two mean seal strengths are statistically different. The difference is said to be statistically insignificant.

However, the confidence interval tells us something more than a *t*-test. It tells us how big (or small) the difference could be. For the seal strength example, the confidence interval shows that the difference in the two mean seal strengths could be as large as 0.9, or the two means could differ by approximately 30 percent of the average seal strength. If a difference of 0.9 is *practically unimportant* to us, then the conclusion is that the difference in the mean seal strength of the two designs is practically unimportant. Only then is a switch to the cheaper design B justified.

What if the difference needs to be less than 0.5 for it to be practically unimportant? Then, based upon the collected data, we cannot make a decision! This is because our analysis shows that while the difference could be as large as 0.9, it could also be as small as 0. We need to know the true difference more precisely before we can make a decision. As you can see from Equation (2.3), we can do this by collecting more data.

As n increases, the width of the confidence interval will shrink and we will be able to say whether the true difference is larger or smaller than 0.5. Note that just going by the results of the *t*-test would have been potentially dangerous in this case, because we would have switched to design B *without realizing the need to collect more data* before making this decision. The consequences could have been unhappy customers and financial losses.

Let us consider this subject of confidence interval versus *t*-test further to see how counterproductive a *t*-test can be in practice. Let Δ represent the *practically important* difference in the two population means. Figure 2.1 shows the various possibilities that may arise.

Let us consider the three cases in Figure 2.1 when the null hypothesis being tested is $H_0: (\mu_A - \mu_B) = 0$. The *t*-test conclusions are shown in Figure 2.1, column (a).

Case A: The confidence interval straddles Δ. The *t*-test conclusion regarding the difference in means being statistically significant or not depends upon whether the confidence interval includes zero or not. The *t*-test, as routinely conducted, will miss the main point that based upon the collected data, we do not know whether the difference is larger or smaller than Δ and therefore, additional data are necessary to make a rational decision. The case of measurement method transfer from R&D to manufacturing, described in the opening paragraphs of this chapter, belonged in this category.

Case B: Confidence interval and the *t*-test come to similar conclusions, but this success of the *t*-test is purely a matter of luck.

Case C: In this case, the difference between the two means is practically unimportant. Both the confidence intervals shown are within $\pm\Delta$. However, for the confidence interval that does not include zero, the *t*-test concludes that the difference is statistically significant. The patch manufacturing case described earlier belonged in this category where the *t*-test identified a small

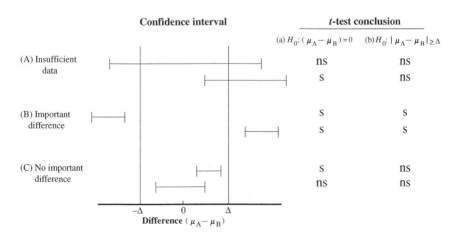

FIGURE 2.1   Confidence interval is better than a *t*-test.

practically unimportant difference to be statistically significant because the sample sizes were very large.

The usual *t*-test for equality of means can be criticized very strongly because the results of the *t*-test can be gamed! If we wish to show equivalence, the strategy would be to collect a small amount of data. Note that from Equation (2.3), the smaller the sample size, the wider the confidence interval. The result will be a very wide confidence interval for $(\mu_A - \mu_B)$ which will include zero and the *t*-test will say that the difference in means is statistically insignificant. If we want to show that the two means are different, then the strategy would be to collect a large amount of data, resulting in a very tight confidence interval for $(\mu_A - \mu_B)$ which will not include zero even if the real difference is very small. The *t*-test will say that the difference is statistically significant. If the confidence intervals are never calculated and shown, perhaps nobody will be the wiser. No wonder statistics sometimes gets a bad name with people thinking that almost anything can be proved by statistics! The saying "there are lies, damn lies and then there are statistics" perhaps comes from such misuses of statistics. The confidence interval analysis is robust. It is not easy to fool people with it.

***Determining* Δ** The determination of Δ, the practically important difference between the two means, is mostly a practical question with some statistical input. We are usually interested in establishing a value of Δ either to find factors that produce important effects or to establish equivalence. Let us consider some examples. Let us suppose that the current yield of a process is 70 percent and we have to increase it to 80 percent. We have one idea in mind, which we have to test in a *t*-test format. Clearly, our objective is not going to be met if the improvement due to this one idea is 2 percent. So Δ could be taken to be (say) 2 percent. More formally, it could be determined on the basis of a return on investment analysis. If the same improvement is to be achieved by using a designed experiment with several factors, one could argue that only a small number of factors and interactions are likely to be important and unless each contributed at least 1 percent yield improvement, we are not likely to meet our objective. Similar practical considerations apply to establishing equivalence. Equivalence does not mean equal. Suppose we have to demonstrate that the measurement method transfer is successful, namely, the results obtained in R&D and manufacturing are equivalent. Let us assume that the specification for the characteristic being measured is 100 ± 15 as is often the case with the drug content of a tablet. A large number of prior R&D experiments suggest that the mean is 99 and the standard deviation is 2. This would imply that most of the R&D data are in the range 93–105. The question in defining equivalence is: If the mean shifts by an amount Δ what will be the practical implication? If the mean shifts by (say) 2 units, a product that is within specification is unlikely to be classified as outside specifications. So, if that is the only practical consideration of importance, then Δ equal to 1 or 2 may be a good choice. On the other hand, if the standard deviation in R&D was 5, even a rather small change in mean would lead to misclassification of products and Δ will have to be very small. Δ should be defined before conducting the experiment. Once Δ is defined, a preliminary sample size calculation (see the next section) should be done to assess the reasonableness of the sample size. Some iterations using the sample size formula may be necessary to achieve a balance between practical values of Δ and the required sample size.

Testing Against Δ One could argue that the previous t-test incorrectly tested the hypothesis $(\mu_A - \mu_B) = 0$ when we were really interested in the hypothesis $|\mu_A - \mu_B| \geq \Delta$. Rejection of the above hypothesis would mean that the difference is practically unimportant. We should have done a t-test against Δ rather than against 0.

Conducting a t-test against Δ is the same as asking whether the confidence interval for $(\mu_A - \mu_B)$ is entirely outside $\pm\Delta$. If so, we would reject the hypothesis that the absolute difference between the two means is less than the practically important difference Δ. The results of such a t-test against Δ can be visually predicted from the confidence intervals for $(\mu_A - \mu_B)$ and are shown in Figure 2.1, column (b). While this t-test correctly identifies the no important difference situation, it continues to misjudge the insufficient data situation.

2.3 HOW MUCH DATA TO COLLECT?

One could then argue that the solution to the insufficient data situation may be found by calculating the correct sample size before data collection. It seems intuitively clear that as the true absolute difference in means gets closer to Δ, the sample size necessary to tell whether the absolute difference in means is smaller than Δ or not will increase and ultimately become infinite. To avoid infinite sample size, we create a *specific* alternate hypothesis and determine the sample size necessary to test the following hypotheses:

$$H_0 : |\mu_A - \mu_B| \geq \Delta$$

$$H_1 : |\mu_A - \mu_B| \leq \delta$$

Note that Δ is the absolute difference in means of practical importance and δ (smaller than Δ) is our guess of what the real but unknown absolute difference may be. In making such decisions, there are two risks of making wrong decisions.

$$\alpha = \text{Probability of rejecting } H_0 \text{ when } H_0 \text{ is true}$$

$$\beta = \text{Probability of rejecting } H_1 \text{ when } H_1 \text{ is true}$$

How much risk should we take? This depends upon the consequences of making wrong decisions. Rejecting H_0 when it is true means switching to the cheaper design when we should not have switched resulting in unhappy customers. Rejecting H_1 when it is true means not switching to the cheaper package when we should have switched implying considerable lost savings. Which risk has the worse consequence? In this case, perhaps it is the α risk. Based upon such considerations, acceptable levels of risks are assigned. Let us assume the risks to be $\alpha = 5$ percent and $\beta = 10$ percent.

This decision-making situation is graphically illustrated in Figure 2.2. The figure shows the distribution of the difference in sample averages under the null and the

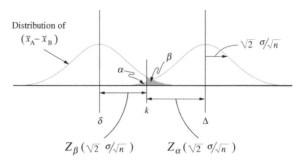

FIGURE 2.2   Calculating sample size.

alternate hypotheses. If the absolute difference between the two observed averages is less than k, the null hypothesis will be rejected, meaning that the probability of rejecting the null hypothesis when it is true is α. Similarly, when the absolute difference between the two averages is greater than k, the alternate hypothesis will be rejected, meaning that the probability of rejecting the alternate hypothesis when it is true is β. With reference to Figure 2.2, change in k increases one risk and reduces the other. As the value of sample size n increases, the standard deviation of $(\bar{x}_A - \bar{x}_B) = \sqrt{2}\sigma/\sqrt{n}$ reduces and the width of the distribution shrinks reducing both risks. Therefore, the desired α and β risks can be obtained by the appropriate choice of n and k.

From Figure 2.2,

$$\Delta - \delta = (Z_\alpha + Z_\beta)\frac{\sigma\sqrt{2}}{\sqrt{n}}$$

This means that when the two standard deviations (for design A and design B) are equal, the sample size is given by

$$n = \frac{2(Z_\alpha + Z_\beta)^2}{d^2} \quad \text{where} \quad d = \frac{\Delta - \delta}{\sigma} \tag{2.4a}$$

When the two standard deviations are different, the more general formula is

$$n = \frac{(\sigma_1^2 + \sigma_2^2)(Z_\alpha + Z_\beta)^2}{(\Delta - \delta)^2} \tag{2.4b}$$

The necessary values of Z_α and Z_β could be obtained from the normal distribution table in the Appendix. The effect of changing the values of α and β risks on the numerator of Equation (2.4a) is shown below. Small risks lead to a large value of numerator for Equation (2.4a) and result in a large sample size for any specified value of d.

α Risk (%)	1	5	10	1	5	10	1	5	10	1	5	10
β Risk (%)	1	1	1	5	5	5	10	10	10	20	20	20
$2(Z_\alpha + Z_\beta)^2$	43	31	26	31	22	17	26	17	13	20	12	9

If the α and β risks are of the order of 5–10 percent, Equation (2.4a) may be approximated by

$$n = \frac{20}{d^2} \quad \text{where} \quad d = \frac{\Delta - \delta}{\sigma} \tag{2.5}$$

It should be intuitively clear that if the variability is large and/or the difference in means to be detected is small, the sample size will be large. This is what the formula says as well. While Equation (2.5) is simple, it is often quite satisfactory in practice. For the package seal strength example, if $\Delta = 0.5$, $\delta = 0.25$, and $\sigma = 0.25$, then $d = 1$ and from Equation (2.5), $n = 20$. We should obtain 20 observations with design A and 20 observations with design B, and then if our prior estimate of σ is correct, we will be able to distinguish between the two hypotheses with the preassigned risks.

The sample size is clearly a function of d. The value of d is difficult to specify perfectly because, for example, σ will only be known within a certain confidence interval based upon past data. One consequence of Equation (2.4) or (2.5) is that the percentage uncertainty in d approximately causes twice as big a percentage change in n. The percentage uncertainty in d has a much larger effect on the numerical value of the sample size when d is small than when it is large. As an example, when $d = 1$, $n = 20$. A 20 percent change in d causes an approximately 40 percent change in sample size causing it to range from 14 to 31. When $d = 0.1$, $n = 2000$. The same 20 percent change in d causes the sample size to range from 1390 to 3125.

The sample size calculation depends upon d, the ratio of $(\Delta - \delta)$ and σ, and not upon their individual values. This is important to keep in mind. I remember a vice president of New Technology saying that statistical sample size calculations did not work in animal experimentation because the variability in these experiments is very large and therefore, the statistically derived sample size will obviously be so large as to be impractical. There are at least two problems with this statement. The first is that a large σ does not necessarily mean a large sample size. Sample size also depends upon the value of $(\Delta - \delta)$. If new technology implies a very large change from the old technology, then the value of $(\Delta - \delta)$ may also be very large. This could make the ratio $(\Delta - \delta)/\sigma$ small enough to make the required sample size manageable.

The second problem with the statement is this. Suppose the calculated sample size is large. What will happen if we arbitrarily decide to use a much smaller sample size to save money and time? It is not uncommon for scientists and engineers in industry to calculate the necessary sample sizes and then have others question the cost of data collection as being too large. The cost of sampling is easy to see; the cost of wrong decisions is often forgotten. There is a way out of this, which is to use Equation (2.4) or (2.5) to back-calculate the value of $(\Delta - \delta)$ for any proposed sample size and see if the back-calculated value of $(\Delta - \delta)$ will be satisfactory for making decisions. For the seal strength example, we calculated the sample size to be 20 corresponding to $(\Delta - \delta) = 0.25$. If $\Delta = 0.5$, it would mean that the sample size of 20 is adequate as long as $\delta \leq 0.25$. What if the sample size was arbitrarily reduced to 5? Then, Equation (2.4) tells us that $(\Delta - \delta)$ will have to be 0.5 for it to be detected as significant. Will this be acceptable? This would imply a value of $\delta = 0$, which is not satisfactory, because we know that δ is likely to be greater than zero. Equations (2.4) and (2.5) not only tell us how to calculate the sample size, but they also allow for a rational dialog in terms of changing the calculated sample size.

The calculated sample size depends upon the specified values of Δ, δ, α, β, and σ. There is really no way to specify these quantities perfectly. This is particularly true of δ. If we could specify the true difference in means δ perfectly, we would not need the experiment! In the statistical calculation of sample size, we specify the problem imperfectly and then solve it perfectly! The Goldilocks sample size cannot be found. Even so, it is important to calculate the sample size as best as can be done to be in the right ballpark.

Calculating Postexperiment Power Power is defined as $100(1 - \beta)$ percent. If the perfect sample size cannot be calculated before collecting the data, then one may argue as follows. It is true that the calculated sample size may not be perfect. However, we can calculate power after the experiment is done. The smaller the calculated postexperiment power, the larger the β risk and if the computed β risk exceeds the desired β risk, it would signal the need for more data. If the original β risk was 10 percent for a desired power of 90 percent, and the postexperiment power turned out to be 70 percent for a β risk of 30 percent, then we know that the sample size was incorrect, more data may have to be obtained before making a decision, and the necessary additional sample size can be calculated.

The postexperiment power may be found by calculating the β risk based upon the observed value of s as shown in Figure 2.3. This figure is different from Figure 2.2 in that σ is replaced by s and the Z-distribution is replaced by the t-distribution. It can be seen by inspection that

$$t_\beta = \frac{(\Delta-\delta)-t_\alpha\left(s\sqrt{2}/\sqrt{n}\right)}{s\sqrt{2}/\sqrt{n}} = (\Delta-\delta)/\left(\sqrt{2}s/\sqrt{n}\right)-t_\alpha \qquad (2.6)$$

Given the predefined values of α, Δ, δ, and the observed n and s, the value of t_β can be calculated. The β risk can be obtained by referring this calculated value of t_β to the t-distribution with $2(n-1)$ degrees of freedom using the t-table in the Appendix.

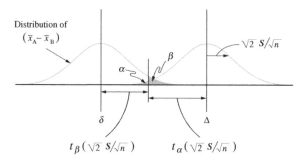

FIGURE 2.3 Calculating postexperiment power $H_0 : |\mu_A-\mu_B| \geq \Delta$.

Let us reconsider the package seal strength example. If $\Delta = 0.5$, $\delta = 0.25$, $\alpha = 5$ percent, $\beta = 10$ percent, and $\sigma = 0.25$, then $d = 1$ and from Equation (2.4), $n = 17$. Suppose 17 observations each were obtained for the two package designs, the observed difference in the two sample averages was 0.2, and the pooled standard deviation turned out to be 0.3. Then, substituting $s = 0.3$ in Equation (2.6), for $\delta = 0.25$, $t_\beta = 0.73$, the corresponding β risk exceeds 25 percent, so the power corresponding to $\delta = 0.25$ is less than 75 percent. This reduction in power is primarily due to the larger observed pooled standard deviation than was originally assumed.

Note that this power calculation only applies to the selected value of δ. If the true absolute difference in means is less than the selected δ, the power to detect the true difference will be higher. If the true absolute difference in means is greater than δ (but less than Δ), the power to detect the true difference will reduce. Since we do not know the true absolute difference in means, δ cannot be perfectly selected. The power calculation will always be imperfect and the problem of calculating the exact additional sample size cannot be solved.

Decisions regarding comparison of means should be made using the confidence interval for difference of means as indicated in Figure 2.1. Presently, the 95 percent confidence interval given by Equation (2.3) is 0.2 ± 0.21. Since this is less than $\Delta = 0.5$, the conclusion is that the difference in means is practically not important. The confidence interval is easy to compute, communicates the results clearly, cannot be gamed, shows the postexperiment power visually, and allows for sequential decision making if required.

2.4   REDUCING SAMPLE SIZE

Data collection is often expensive. It is, therefore, important to consider ways to reduce the sample size necessary to make reasoned decisions. In this chapter, we have so far considered the sample size necessary to compare the means of two normally distributed populations with the necessary sample size given by Equation (2.4). However, in general, comparative experiments may involve not only comparison of means, but also comparison of variances and fraction defectives based upon attribute data. Additionally, we may be interested in comparing multiple populations, not just two populations. These situations require sample sizes greater than that given by Equation (2.4). The necessary sample size formulae are given in Joglekar (2003). From a practical viewpoint, one rule of thumb is useful to remember—it generally requires a double-digit sample size to compare means, a triple-digit sample size to compare variances, and a four-digit sample size to compare small changes in fraction defective.

Here are some ways to reduce the sample size:

1. With reference to Equation (2.4), the sample size can be reduced substantially if the value of d is made larger. This can be done in three ways—by reducing σ, increasing Δ, and reducing δ. The value of σ is based upon our estimate of standard deviation and cannot be reduced unless the variability itself is reduced

by making improvements. Reducing the value of σ is not a practical suggestion at the time of estimating the sample size. Δ defines the practically important difference in the two population means. The tendency to select an unnecessarily small value for Δ should be resisted, particularly if the calculated sample size is too large. δ is our prior estimate of the true difference in the two population means. This is an unknown and, therefore, a smaller value could be selected so long as we are willing to collect additional data, if required, to make a decision.

2. With reference to Equation (2.4), a smaller sample size will result if the α and β risks are increased. Arbitrarily large values of risks should not be chosen. On the other hand, while it is a common practice to work with 95 percent confidence levels implying a 5 percent risk, in many situations, 90 percent confidence may be more than sufficient. Risks need to reflect the consequences of being wrong.

3. As explained in Chapter 3, a well-designed experiment leads to large reductions in effective sample sizes when the effects of multiple factors are to be determined. As an example, suppose we have to determine the effects of six factors. If we change these six factors from some base point, the minimum one factor at a time experiment will involve seven trials. For $d = 1$, Equation (2.5) will lead to a total of 140 trials. Instead, for a well-designed fractional factorial experiment, the total number of trials will be approximately 64, which is a very large reduction in the cost of data collection.

4. Replacing attribute data by variable data will lead to a large reduction in sample size. Collecting data on a continuous scale, recording it as attribute data by noting whether the individual values were within or outside the specification, and then making attribute comparisons is often a particularly bad idea. There are many situations where the attribute data cannot be replaced by variable data, or such a replacement is very expensive. In such situations, worst-case testing should be attempted to reduce the sample size. This may be understood as follows. Suppose we have to assess the breakability of a glass bottle and a steel bottle assuming for the moment that we do not understand the difference between glass and steel *a priori*. If both are dropped from a small height on to a carpeted floor, it will take a very large number of drops to demonstrate that the glass bottle breaks and the steel bottle does not break. On the other hand, if we drop the bottles from a considerable height on to a cement floor, a few drops will prove the point.

5. As explained in Chapter 4, the problem of comparing variances can effectively be converted into the problem of comparing means by the use of noise factors. This can dramatically reduce the sample size because we do not depend upon random replication to assess variability, but effectively induce it by changes in the noise factor. Such a dramatic reduction in sample size is possible only if the noise factor, namely, the cause of variability, is known and can be experimented with. Otherwise, large sample sizes are necessary to detect small changes in variability as explained below in the section on the F-test (also see Joglekar (2003)).

2.5   PAIRED COMPARISON

Let us now consider another example involving the analysis of capital market returns discussed in Chapter 1. Table 2.2 shows the yearly returns of small company stocks and large company stocks over the 80-year period from 1924 to 2005 (see Ibbotson (2006)). We want to answer a question posed in Chapter 1: Is the mean yearly return for small company stocks greater than the mean yearly return for large company stocks and if so, by how much?

The percent yearly returns in Table 2.2 can be analyzed to obtain the averages and standard deviations of returns for the small and large company stocks. The computed values are $\bar{x}_{small} = 17.4$, $s_{small} = 32.9$, $\bar{x}_{large} = 12.3$, and $s_{large} = 20.2$.

The F-test and the confidence interval for the ratio of two standard deviations (see the next section and also Joglekar (2003)) show that the small company stocks have significantly larger standard deviation than the large company stocks. Thus, the risk is higher for the small company stocks.

There are two ways to determine how big the difference is between the average yearly returns from the small and large company stocks. One approach is to do an independent comparison (called the independent t-test) and the other is to do a paired comparison (called the paired t-test).

Independent Comparison   We can construct a confidence interval for difference of means, as we have done before, using a slightly generalized version of Equation (2.3) to account for the fact that the two standard deviations are different. The 95 percent confidence interval for $(\mu_{small} - \mu_{large})$ is given by

$$\left(\bar{x}_{small} - \bar{x}_{large} \right) \pm t_{n_1 + n_2 - 2, 0.025} \sqrt{\frac{s^2_{small}}{n_1} + \frac{s^2_{large}}{n_2}} \qquad (2.7)$$

where n_1 and n_2 are the two sample sizes, each equal to 80 in this case. Note that when the two sample sizes are equal, Equation (2.7) becomes the same as Equation (2.3). Upon substitution, the 95 percent confidence interval for the difference in means turns out to be -3.5 to 13.7. Since the confidence interval includes zero, we cannot say that the mean yearly return for the small company stocks is statistically larger than that for the large company stocks. Had we done an independent t-test, the result would have been statistically insignificant.

Paired Comparison   The very wide confidence interval calculated above is due to the large standard deviation of returns. This large standard deviation of returns results from the changing economic conditions from one year to the next. A more precise comparison of the mean yearly returns can be made by pairing the returns by year, namely, by calculating the difference between small and large company stock returns for each year. The results are shown in Table 2.2. The difference column in Table 2.2 is not affected by the year-to-year variability of returns and thereby allows a more precise comparison. This approach to making comparisons is known as paired comparison and the approach can only be used when there is a meaningful way to pair observations.

TABLE 2.2 Percent Yearly Returns of Small and Large Company Stocks (1926–2005)

Year	Small	Large	Difference (Large-Small)
1926	11.62	0.28	−11.34
1927	37.49	22.10	−15.39
1928	43.61	39.69	−3.92
1929	−8.42	−51.36	−42.94
1930	−24.90	−38.15	−13.25
1931	−43.34	−49.75	−6.41
1932	−8.19	−5.39	2.80
1933	53.99	142.87	88.88
1934	−1.44	24.22	25.66
1935	47.67	40.19	−7.48
1936	33.92	64.80	30.88
1937	−35.03	−58.01	−22.98
1938	31.12	32.80	1.68
1939	−0.41	0.35	0.76
1940	−9.78	−5.16	4.62
1941	−11.59	−9.00	2.59
1942	20.34	44.51	24.17
1943	25.90	88.37	62.47
1944	19.75	53.72	33.97
1945	36.44	73.61	37.17
1946	−8.07	−11.63	−3.56
1947	5.71	0.92	−4.79
1948	5.50	−2.11	−7.61
1949	18.79	19.75	0.96
1950	31.71	38.75	7.04
1951	24.02	7.80	−16.22
1952	18.37	3.03	−15.34
1953	−0.99	−6.49	−5.50
1954	52.62	60.58	7.96
1955	31.56	20.44	−11.12
1956	6.56	4.28	−2.28
1957	−10.78	−14.57	−3.79
1958	43.36	64.89	21.53
1959	11.95	16.40	4.45
1960	0.47	−3.29	−3.76
1961	26.89	32.09	5.20
1962	−8.73	−11.90	−3.17
1963	22.80	23.57	0.77
1964	16.48	23.52	7.04
1965	12.45	41.75	29.30
1966	−10.06	−7.01	3.05
1967	23.98	83.57	59.59
1968	11.06	35.97	24.91
1969	−8.50	−25.05	−16.55

TABLE 2.2 (*Continued*)

Year	Small	Large	Difference (Large-Small)
1970	4.01	−17.43	−21.44
1971	14.31	16.50	2.19
1972	18.98	4.43	−14.55
1973	−14.66	−30.90	−16.24
1974	−26.47	−19.95	6.52
1975	37.20	52.82	15.62
1976	23.84	57.38	33.54
1977	−7.18	25.38	32.56
1978	6.56	23.46	16.90
1979	18.44	43.46	25.02
1980	32.42	39.88	7.46
1981	−4.91	13.88	18.79
1982	21.41	28.01	6.60
1983	22.51	39.67	17.16
1984	6.27	−6.67	−12.94
1985	32.16	24.66	−7.50
1986	18.47	6.85	−11.62
1987	5.23	−9.30	−14.53
1988	16.81	22.87	6.06
1989	31.49	10.18	−21.31
1990	−3.17	−21.56	−18.39
1991	30.55	44.63	14.08
1992	7.67	23.35	15.68
1993	9.99	20.98	10.99
1994	1.31	3.11	1.80
1995	37.43	34.46	−2.97
1996	23.07	17.62	−5.45
1997	33.36	22.78	−10.58
1998	28.58	−7.31	−35.89
1999	21.04	29.79	8.75
2000	−9.11	−3.59	5.52
2001	−11.88	22.77	34.65
2002	−22.10	−13.28	8.82
2003	28.70	60.70	32.00
2004	10.87	18.39	7.52
2005	4.91	5.69	0.78

The calculated values of the average difference and the standard deviation of difference are $\bar{x}_{\text{difference}} = 5.07$ and $s_{\text{difference}} = 21.1$. We can now calculate a simple 95 percent confidence interval for $\mu_{\text{difference}}$ given by

$$\bar{x}_{\text{difference}} \pm t_{n-1,0.025}\left(\frac{s_{\text{difference}}}{\sqrt{n}}\right) \tag{2.8}$$

Upon substitution, the 95 percent confidence interval for the difference of means is (0.37–9.77). Note that by removing the year-to-year variability, the confidence interval has become tighter. Since this confidence interval does not include zero, the conclusion is that we are 95 percent sure that the mean yearly return for the small company stocks is statistically larger than that for the large company stocks. Had we done a paired *t*-test, it would also have identified the difference to be statistically significant. The confidence interval shows something more than the *t*-test: that the mean yearly return for small company stocks could be larger than that for the large company stocks by as small as 0.37 percent or as large as 9.77 percent. This confidence interval is tighter than the one calculated using the independent *t*-test, but continues to be wide because the standard deviation of the difference between yearly returns of small and large company stocks is very large.

2.6   COMPARING TWO STANDARD DEVIATIONS

Just as a *t*-test is used to compare two population means, an *F*-test is used to compare two population standard deviations. Even when comparing two means, the *F*-test should be done first because the *t*-test can be conducted assuming the two standard deviations to be equal or unequal. If the two standard deviations can be assumed to be equal, the resultant *t*-test is more sensitive and the confidence interval is narrower. Thus, the reasonableness of the assumption of equality of standard deviations needs to checked first. This could be remembered as follows: Alphabetically *F* comes before *t*!

To test the hypothesis $\sigma_1 = \sigma_2$ versus $\sigma_1 \neq \sigma_2$, the ratio

$$F = \frac{S_1^2/\sigma_1^2}{S_2^2/\sigma_2^2} \qquad (2.9)$$

is used. This ratio has an *F*-distribution, which is a skewed distribution characterized by the degrees of freedom used to estimate S_1 and S_2, respectively called the numerator degrees of freedom $(n_1 - 1)$ and denominator degrees of freedom $(n_2 - 1)$. Under the null hypothesis $\sigma_1 = \sigma_2$, the *F*-ratio becomes S_1^2/S_2^2. In calculating the *F*-ratio, the larger variance is conventionally put in the numerator so that the calculated value of *F* is always greater than one. If the computed value of *F* exceeds the critical value $F_{\alpha/2, n_1 - 1, n_2 - 1}$ for a two-sided alternate hypothesis or $F_{\alpha, n_1 - 1, n_2 - 1}$ for a one-sided alternate hypothesis, the null hypothesis is rejected. These critical values are tabulated in the Appendix.

Let us consider an example involving sporting ammunition. Precision is important for sporting ammunition. For a certain design of ammunition, a total of 21 shots were fired and the standard deviation of the resulting scatter was determined. The results were $n_1 = 21$ and $s_1 = 0.8$. An improved design resulted in $n_2 = 16$ and $s_2 = 0.6$. Has the improved design reduced variability? If σ_1 and σ_2 denote the population standard deviations for the old and the new designs, then the null hypothesis is $\sigma_1 \leq \sigma_2$. The alternate hypothesis is $\sigma_1 > \sigma_2$. The computed value of the *F*-statistic is

$$F = (0.8/0.6)^2 = 1.78$$

The critical F value for $\alpha = 5$ percent is $F_{0.05,20,15} = 2.33$. Since the computed F value is less than 2.33, the null hypothesis cannot be rejected, namely, there is no conclusive statistical evidence that the new design has reduced variability.

Just as the t-test should be replaced by the confidence interval for difference of means, the F-test should be replaced by the confidence interval for the ratio of variances. The $100(1 - \alpha)$ percent confidence interval for σ_1^2/σ_2^2 is

$$\frac{\left(s_1^2/s_2^2\right)}{F_{\alpha/2,n_2-1,n_1-1}} \text{ to } \left(s_1^2/s_2^2\right)F_{\alpha/2,n_1-1,n_2-1} \tag{2.10}$$

Confidence interval for σ_1/σ_2 is computed by taking square roots. For the ammunition example, $n_1 = 21$, $n_2 = 16$, and for $\alpha = 10$ percent (i.e., 90 percent confidence), the confidence interval for σ_1^2/σ_2^2 is

$$\frac{\left(0.8^2/0.6^2\right)}{2.2} \text{ to } \left(0.8^2/0.6^2\right) \times 2.33 = 0.8 \text{ to } 4.14$$

The 90 percent confidence interval for σ_1/σ_2 is 0.89–2.03, obtained by taking square roots. The fact that the confidence interval for σ_1/σ_2 includes 1 is the same as saying that the F-test result would have been statistically insignificant. Thus, the confidence interval and the F-test come to the same conclusion, namely, that there is no conclusive statistical evidence that the new design has reduced variability. The confidence interval provides us with more information than the F-test. It says that the standard deviation of the old design could be 10 percent smaller or 100 percent larger than the standard deviation of the new design! This confidence interval is very wide. We are not even able to say which design is better, let alone by how much. The sample sizes are insufficient to make a reasonable decision.

The sample sizes necessary to detect a certain ratio of two standard deviations for an α risk of 5 percent and a β risk of 10 percent are given in Table 2.3 for various values of σ_1/σ_2.

It can be seen that to detect a standard deviation ratio of less than 1.4, more than 100 observations per population are required. For the ammunition example, if the new design reduces the standard deviation by 30 percent compared with the prior design, it would be considered a significant improvement. What should the sample size be? The

TABLE 2.3 Sample Size to Detect Changes in Standard Deviation

Ratio of Standard Deviations to Be Detected	Sample Size
3.0 to 1	9
2.5 to 1	13
2.0 to 1	20
1.75 to 1	30
1.5 to 1	50
1.4 to 1	100

standard deviation ratio is $1/0.7 \approx 1.4$ and the sample size is 100. A nomogram to determine sample size necessary to detect any desired ratio of standard deviations is given in Joglekar (2003).

2.7 RECOMMENDED DESIGN AND ANALYSIS PROCEDURE

We need to use a robust analysis procedure for hypothesis testing. For example, to compare two means, we seek a procedure that does not penalize us heavily because we are unable to specify the problem (i.e., $\Delta, \delta, \alpha, \beta$, and σ) perfectly. That procedure is the following:

1. Correctly define the null and the alternate hypotheses to suit the specific decision to be made.
2. Think through what the selected values of Δ, δ, α, β, and σ should be. This thinking is critical to make good decisions even when the selected values are imperfect.
3. Compute the necessary sample size based upon the selected values of $\Delta, \delta, \alpha, \beta$, and σ. This sample size will not be perfect, but is the best that can be done.
4. Conduct the experiment and collect data.
5. Forget the *t*-test! This is particularly so if comparisons are being made using available data without a proper computation of the sample size.
6. Calculate the confidence interval for the difference in population means and decide as shown in Figure 2.1. Figure 2.1 is easy to compute, communicates the results clearly, cannot be gamed, shows the postexperiment power visually, and allows for sequential decision-making if required.

The above arguments apply equally well to all *t*-tests, *F*-test, tests for normality, and all usual hypothesis tests. The implication is that these tests should be replaced by their confidence interval analogs for meaningful decision-making.

2.8 QUESTIONS TO ASK

Asking the following questions will help improve hypothesis test based decision-making.

1. Do the standard operating procedures in your company use the usual *t*-test to make decisions? If so, these procedures should be reviewed and replaced by the confidence interval based approach described in this chapter.
2. For continuous data, are the assumptions of normality, independence, and constant variance being checked, and are they satisfied? For discrete data, are the binomial or Poisson assumptions satisfied?

3. Do people in your company compute a sample size before conducting the t-test? If not, sample size computations should be encouraged.

4. How are the values Δ, δ, α, β, and σ determined when computing the sample size? The value of Δ is based upon both practical and statistical considerations, δ is an educated guess, and σ is based upon past history and experience. The α and β risks are based upon business consequences of wrong decisions. Do you agree with how these numbers are being established in your company?

5. What attempts are being made to reduce the need for large sample sizes? Are ideas such as replacing discrete measurements by continuous measurements, accelerated testing, worst-case testing, and multifactor designed experiments being considered?

6. How are the data being analyzed to make a decision? After computing the sample size, the practice of conducting a t-test to make decisions should be discouraged. The use of confidence interval analysis procedure should be encouraged with decisions made as shown in Figure 2.1.

7. If people in your company routinely do a t-test with any available data with arbitrary sample sizes, it is even more important that you ask them to show the result as a confidence interval. Is the confidence interval being correctly interpreted in making decisions? What decisions will they make after looking at the confidence interval?

8. For every decision made, which of the following conclusions was reached: Meaningful difference, no meaningful difference, or collect more data?

9. It should not escape your attention that the above concerns and solutions apply not only to all types of t-tests (single-sided, double-sided, paired, etc.), but also to F-tests, normality tests, and all hypothesis tests. All usual hypothesis tests suffer from the same criticism that the usual t-test does and need to be replaced by suitable confidence interval procedures.

Further Reading   The book by Joglekar (2003) discusses the commonly used hypothesis tests such as tests to compare means, standard deviations, and proportions for single-population, two-population, and multiple-population situations. These tests include the various Z-tests, t-tests, F-tests, and ANOVA. It also provides the confidence intervals for some of the hypothesis tests. Formulae to compute sample sizes are also given. The book by Box et al. (1978) also has a discussion on significance tests and confidence intervals. Ibbotson (2006) provides the capital market data.

CHAPTER 3

DESIGN OF EXPERIMENTS: IS IT NOT GOING TO COST TOO MUCH AND TAKE TOO LONG?

There continues to be an insufficient use of design of experiments (DOE) in industry and a misconception that DOE takes too long and costs too much. The purpose of DOE is to cost-effectively accelerate experiment-based learning and thereby accelerate the research and development of products and processes. DOE accomplishes this through the planning of effective and efficient experiments and a robust analysis procedure.

Gold Medal Ammunition   In 1991, a year before the Barcelona Olympics, I received a call from a sporting ammunition company to provide training and consultation on design of experiments and statistical process control. This company had set the goal to have the U.S. Olympic shooting team use U.S.-made ammunition at the Barcelona Olympics. Before 1992, British ammunition was the ammunition of choice at the Olympic Games because of its precision. An engineering team had been set up to develop and manufacture the ammunition. The team was trained in the use of design of experiments and statistical process control. They were provided with guidance in the correct implementation of these methods. In a span of less than a year, the team designed the ammunition and its manufacturing process such that the precision of this new ammunition surpassed that of the British ammunition. In pre-Olympic trials, the U.S. Olympic team preferred this ammunition. As luck would have it, United States won the gold medal at Barcelona. The company marketed the ammunition as the gold-medal ammo. Much can be accomplished with leadership, the right tools, and some luck.

Industrial Statistics: Practical Methods and Guidance for Improved Performance   By Anand M. Joglekar
Copyright © 2010 John Wiley & Sons, Inc.

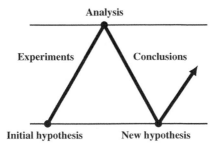

FIGURE 3.1 The learning cycle.

The Learning Cycle R&D may be conceptually viewed as a learning cycle as shown in Figure 3.1. Suppose we have to develop a new drug formulation. The cycle begins with an initial hypothesis regarding the formulation components to use and their proportions. This initial hypothesis is based upon our theoretical and practical understanding of the subject matter. We realize that we do not know everything that we need to know and therefore must experiment to find the necessary answers. As shown in Figure 3.1, experiments are conducted, data are gathered and analyzed, conclusions are drawn, and this leads to a new hypothesis regarding how to formulate the product. New ideas and questions arise, and the cycles continue. At some point, the manufacturing process needs to be developed, scaled up, validated, transferred to manufacturing, and improved. The cycles continue. How can this cyclic process be accelerated?

One possibility is to shorten the length of time it takes for each cycle. Much of the time is taken up by experimentation and data collection. So the way to reduce the cycle time is to reduce the number of experiments necessary to generate relevant knowledge. A second possibility is to reduce the number of cycles necessary to complete the project. There are two reasons for the large number of cycles. If some of the conclusions drawn at the end of each cycle are incorrect, additional cycles are necessary. Also, if experiments are planned without a proper consideration of available technical knowledge (see Chapter 10) and downstream issues, a much larger amount of effort is required later. DOE is the methodology to develop well-planned experiments and to analyze them to draw correct conclusions.

3.1 WHY DESIGN EXPERIMENTS?

The traditional approach to experimentation is to conduct one-factor-at-a-time (OFAAT) experiments. This approach can result in wrong conclusions leading to suboptimization while requiring larger experimental effort as explained in Box et al. (1978). This may be seen as follows.

Let us suppose that two factors, time and temperature, influence the yield of a process. We have to optimize the process by finding the values of time and temperature that maximize yield. We believe in OFAAT. Our initial judgment is that a temperature of 255 degrees is reasonable, so we fix temperature at 255 degrees and vary time from

5 to 25 minutes as shown in Figure 3.2(a). The optimum time at a temperature of 255 degrees is 16.5 minutes. We are not completely sure whether the temperature of 255 degrees is the best, so we have to experiment with temperature. Where to keep the time constant as we vary temperature? A time of 16.5 minutes seems to be a reasonable choice since it is the best time found so far. Figure 3.2(b) shows the results of the second set of experiments. The temperature of 255 degrees indeed appears to be the best.

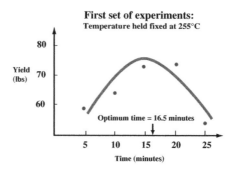

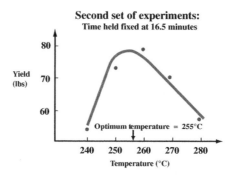

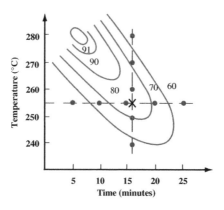

FIGURE 3.2   (a–c) Suboptimization.

The conclusion is that a time of 16.5 minutes and a temperature of 255 degrees lead to the largest possible yield.

The fact that this conclusion is wrong can be seen from Figure 3.2(c). For any combination of time and temperature, there is a yield. We can think of yield as a hill in the time–temperature domain. At the start of the experimentation, this is an invisible hill. The hill may be as shown by the contours of constant yield in Figure 3.2 (c). For any combination of time and temperature on the contour labeled 80, the yield is 80 pounds. A maximum yield of over 91 pounds occurs at a time of 5 minutes and a temperature of 280 degrees. Our OFAAT experiments missed this optimum. Given the shape of this hill (not all hills are alike) it is easy to see from Figure 3.2(c) how our OFAAT experiments led us astray to a suboptimal yield of 75 pounds. The fact that our optimum takes three times as long as it should implies a going-out-of-business curve in a competitive world. The conclusions drawn at the end of these OFAAT cycles are wrong because the experiment was poorly designed.

An important reason for these wrong conclusions is the inability of OFAAT experiments to determine the interactions. Two factors are said to interact when the effect of one factor depends upon the level of another. A lack of consistent effect is an indication of interaction. This can be easily seen from the familiar situation depicted in Figure 3.3. In the first experiment (0, 0), the base point where there are no male and female rabbits present, no babies are born. The same result is obtained when there are male rabbits without female rabbits or when there are female rabbits without male rabbits. This is what the OFAAT experiment would be, varying one factor at a time from the base point experiment. It is only when both male and female rabbits are together that the babies are born. Clearly, the effect of male rabbits depends upon the level (number) of female rabbits, that is, male and female rabbits *interact* in the sense of the design of experiments use of the word interaction. OFAAT experiments cannot determine such interactions.

OFAAT experiments are inefficient. Suppose we have to find the effects of time and temperature on yield as shown in Figure 3.4(a). At each combination of time and temperature, four replicated experiments are conducted because the replication error is large. So by conducting 12 OFAAT experiments, we can find the effects of time and

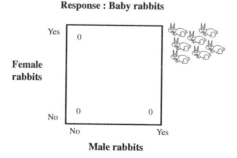

FIGURE 3.3   Anatomy of an interaction.

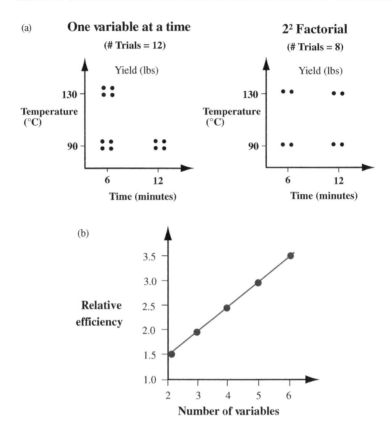

FIGURE 3.4   (a) OFAAT is inefficient and (b) DOE is efficient.

temperature. Instead, we could have done four trials (known as a factorial experiment) as shown in Figure 3.4(a) where an extra combination of time and temperature is added. Suppose we have only two replicates at each combination of time and temperature. To determine the effect of time, we compare the average yield of four trials at high time with the average yield of the four trials at low time, a four versus four comparison as with the OFAAT experiment. The same is true for temperature. We can also determine the effect of the interaction between time and temperature because we can estimate the effect of time at two different temperatures and find out if it varies. With a factorial design strategy, we obtain more information with fewer experiments. This is more than a free lunch! Figure 3.4(b) shows the relative efficiency of DOE as a function of the number of factors in the experiment. Relative efficiency is defined as the ratio of the number of trials using OFAAT divided by the number of trials using DOE necessary to obtain equivalent information. The figure shows that with five factors, OFAAT requires three times as many trials as DOE does; in this case, we can reduce the time and cost of experimentation by a factor of three using DOE.

DOE Strategy   The classical approach to design of experiments is a two-step strategy—screening followed by optimization. In designing a new product or a new process, we are interested in determining the effects of a large number of design factors. The Pareto principle or the law of vital few and trivial many applies. We have to determine the key design (i.e., control) factors that produce large effects and this is accomplished by conducting screening experiments. The purpose of screening experiments is twofold: to identify the key control factors and interactions under the assumption of linear effects and to test this assumption of linearity. If the linearity assumption turns out to be true, the screening experiment itself is sufficient to optimize the product and process design. If the linearity assumption fails, then, to optimize product and process designs, additional optimization experiments are necessary to determine the factors that produce curvilinear effects. Optimization experiments necessarily require three levels for each factor to estimate curvature. This is why the overall optimization problem is broken up into a two-step strategy—screening followed by optimization—because if we design an optimization experiment without knowing which factors are important, the experiment will become unmanageably large.

3.2   FACTORIAL DESIGNS

Experiment Design   Factorial design is a screening experiment. The strategy is simple—it is to select the factors whose effects we have to investigate, decide the number of levels for each factor, and then conduct all possible combinations of the levels of all factors. If there are three factors, each at two levels, then there are $2 \times 2 \times 2 = 8$ combinations of levels and such an experiment is called a 2^3 factorial design. Designing a factorial experiment is trivial. Consider an experiment discussed in Box et al. (1978) involving three control factors: temperature (T), concentration (C), and catalyst (K). The first two are continuous factors while catalyst is a discrete factor. The response is yield measured in pounds per batch. We are interested in determining the effects of these factors on the yield of the chemical process. On the basis of engineering knowledge, the ranges selected for each factor over which it is sensible to experiment are temperature (160–180 degrees), concentration (20–40 percent), and catalyst (A and B). In terms of coded units, the low level of each factor is denoted by -1 and the high level is denoted by $+1$. If we experiment at only two levels of each factor, the eight combinations in coded units are easy to write as shown in Table 3.1. The first column goes minus, plus, the second goes two minuses and two pluses, and so on. This is just an easy way to write all possible combinations. The coded experiment can then be translated to original units as shown. Designing a factorial experiment is as simple as this. We obtain the values of yield by conducting the experiment.

Graphical Analysis   The analysis of the factorial experiment is also simple. One way is to analyze it graphically. This experiment is represented as a cube in Figure 3.5.

TABLE 3.1   Factorial Design

	Original Units			Coded Units			Yield
Trials	T	C	K	T	C	K	Y
1	160	20	A	−	−	−	60
2	180	20	A	+	−	−	72
3	160	40	A	−	+	−	54
4	180	40	A	+	+	−	68
5	160	20	B	−	−	+	52
6	180	20	B	+	−	+	83
7	160	40	B	−	+	+	45
8	180	40	B	+	+	+	80

The cube leads to the following conclusions:

1. Temperature has a large positive effect. The right-hand face of the cube has a larger yield than the left-hand face. Positive means that as temperature increases so does the yield.
2. Concentration has a small negative effect. The top face of the cube has a smaller yield than the bottom face. As concentration increases, yield reduces.
3. There is a temperature–catalyst (TK) interaction. The effect of temperature is small with catalyst A (front face of the cube) and large with catalyst B (back face of the cube). Another way to view this interaction is to realize that catalyst A is better at a lower temperature and catalyst B is better at a higher temperature. Which catalyst is better is a function of temperature.

One purpose of conducting the experiment was to identify the key factors and interactions, and this purpose is essentially met by a visual examination. It is good to develop this habit of visual interpretation.

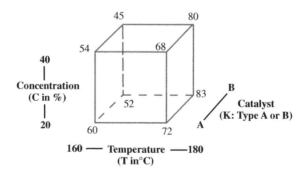

FIGURE 3.5   Graphical representation of the factorial experiment.

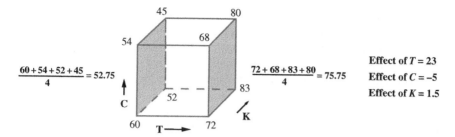

$$\frac{60+54+52+45}{4}=52.75$$

$$\frac{72+68+83+80}{4}=75.75$$

Effect of $T = 23$

Effect of $C = -5$

Effect of $K = 1.5$

FIGURE 3.6 Main effect as a difference between two averages.

Quantifying Effects We can quantify these effects as follows. The main effect of a factor is defined to be the average change in the response as the factor is changed from its low level to its high level. As shown in Figure 3.6, the main effect of temperature is $+23$ (75.75 − 52.75). The main effect of concentration is −5 and the main effect of catalyst is 1.5. These are *average effects*, that is, as temperature increases from 160 to 180 degrees, the yield increases by 23 pounds, *on the average*, not from one corner point to another.

The interaction effect may be computed as follows. Figure 3.7 shows the interaction between temperature and catalyst. With catalyst A, the effect of temperature is $+13$ (from the front face of the cube) and with catalyst B, the effect of temperature is $+33$ (from the back face of the cube). The average effect of temperature is the average of these two numbers, namely, $+23$, as calculated before. This average effect of temperature does not stay fixed, it changes as a function of the catalyst and, therefore, we say that temperature and catalyst interact. The value by which the main effect of temperature changes as a function of catalyst is a measure of the interaction between these two factors. The interaction is $+10$. The effect of temperature becomes 33(23 + 10) or 13(23 − 10) as a function of the catalyst used.

So far, we have conducted essentially a graphical analysis of the factorial experiment using the cube as a reference. The visualization becomes difficult beyond three factors. Also, we have to computerize computations. A simple arithmetic approach to determining effects is shown in Table 3.2, and it will work for any number of factors.

The first three columns of the matrix are the designed experiment and are referred to as the design matrix. While analyzing the experiment, we are also interested in

		Temperature (T)	
		160	**180**
Catalyst (K)	**A**	(54, 60) 57	(68, 72) 70
	B	(45, 52)	(80, 83) 81.5

13 → , 23 , 33 →

$$\text{TK interaction} = \frac{33-13}{2} = 10$$

FIGURE 3.7 Calculation of interaction.

determining the interactions. The interaction columns of the matrix are created by simple multiplication of the corresponding factor columns. For example, the temperature–catalyst interaction column, represented by TK, is obtained by multiplying (dot product) the T and K columns. The entire matrix is referred to as the analysis matrix.

As shown in Equation (3.1), if we take the sign structure for column T, attach it to the data and divide by half the number of trials, the answer is 23, which is the main effect of temperature. If this procedure always worked, we could find all effects by simply repeating it for each column. Why does it work? Because the computation shown in Equation (3.1) for the effect of T is nothing but the difference between the two averages, the average yield at high temperature minus the average yield at low temperature. This is the definition of the main effect of temperature. The fact that the simple computation works is not a surprise.

$$\text{Effect of } T = \frac{-60 + 72 - 54 + 68 - 52 + 83 - 45 + 80}{4} = 23 \qquad (3.1)$$

The calculated main effect of catalyst is 1.5. How are we so sure that the effect is in fact due to the catalyst and not due to any other factor or interaction in the experiment? After all, we have changed multiple factors simultaneously; this is not an OFAAT experiment. We have been told from childhood not to change many factors together because we will not be able to figure out which effect was caused by which factor. This is not a concern with a well-designed experiment because the experiment is a balanced (orthogonal) experiment. It is easy to see this for catalyst (factor K) as follows.

TABLE 3.2   A Simple Method to Find Effects

Trial	T	C	K	TC	TK	CK	TCK	Y
1	−	−	−	+	+	+	−	60
2	+	−	−	−	−	+	+	72
3	−	+	−	−	+	−	+	54
4	+	+	−	+	−	−	−	68
5	−	−	+	+	−	−	+	52
6	+	−	+	−	+	−	−	83
7	−	+	+	−	−	+	−	45
8	+	+	+	+	+	+	+	80

Calculated Effects

Effect	Estimate
T	23.0
C	−5.0
K	1.5
TC	1.5
TK	10.0
CK	0.0
TCK	0.5

From Table 3.2, the effect of factor K is determined by taking the average yield for the last four trials and subtracting from it the average yield for the first four trials. When we take these two averages, in each case, all other factors and interactions appear twice at the plus level and twice at the minus level. Thus, when we compute each average, the effects of all other factors and interactions are cancelled out. This property of balance is true not only for factor K, but also for every other factor and interaction column. A well-designed experiment is a balanced experiment.

Determining Significance The calculated main and interaction effects are shown in Table 3.2. The question now is: Which of these effects are significant? It is well known that not all factors and interactions will produce equal effects; some effects will be much larger than the others. This idea, "the law of the vital few and the trivial many" is also known as the Pareto principle. It is the purpose of a screening experiment to identify these vital few significant effects. There are two kinds of significance: statistical significance and practical significance. An effect is said to be statistically significant if on the basis of data it could be said that the probability of the effect being zero is small. An effect is practically significant if it is large enough to be of practical utility.

Statistical significance can be determined either by constructing a confidence interval for effects or by determining the level of significance using a hypothesis test approach. As explained in Chapter 2, the confidence interval approach is better. This analysis is usually delegated to a software package. The results for significant factors and interactions are shown in Table 3.3. None of the confidence intervals include zero except for factor K (catalyst). Even though the confidence interval for K includes zero, K is considered to be a statistically significant factor because it is involved in a statistically significant interaction.

The practical significance of effects may be judged by expressing the effect as a percent of the average yield. The average yield is 64 pounds. The practical significance column shows that we can use the main effect of temperature to make a 36 percent change in yield compared with the average. This is a large change for the problem at hand and, therefore, the effect of temperature is a practically important effect. On the other hand, if the objective was to make a 10-fold increase in yield, a 36 percent change may not be considered to be practically important, even while the effect was statistically significant. To be important, the effect should be practically and statistically significant.

TABLE 3.3 Significance of Effects

Effect	Statistical Significance: 95% Confidence Interval	Practical Significance: Effect as % of Mean
Mean	64.25 ± 2.9	
T	23.0 ± 2.9	36
C	-5.0 ± 2.9	-8
K	1.5 ± 2.9	2
TK	10 ± 2.9	16

Building Models The estimated effects immediately lead to an equation for yield as a function of the important factors and interactions. The equation is also called a model.

$$\text{Yield} = 64.25 + 11.5 \text{ temperature} - 2.5 \text{ concentration} + 0.75 \text{ catalyst}$$

$$+ 5.0 \text{ temperature} \times \text{catalyst} + \text{residual} \qquad (3.2)$$

This equation is written in the coded units where each factor increases from -1 to $+1$. An effect corresponds to a change of two coded units in the factor and a coefficient corresponds to a change of one unit. Therefore, each coefficient in the equation is half the effect. The equation can also be rewritten in original units as necessary. Residual is the portion of yield that is not explained by the equation. The goodness of the equation can be assessed from the standard deviation of the residual, which in this case is 1.3, that is, the predicted yield could deviate from the true value by approximately ± 2.6 pounds. Software packages readily provide this information. The larger this prediction error, the less useful the equation becomes. Large prediction errors will also suggest the possibility that some important factors are not included in the equation (and in the original experiment) or that the measurement error is very large.

In summary, a factorial experiment, and in general a screening experiment, has the following objectives:

1. To identify important factors and interactions. In this case, the main effects of temperature, concentration, and catalyst, and the temperature–catalyst interaction are important.
2. To determine a functional relationship between the response and the identified important factors and interactions, and to assess whether the equation is adequate for the purpose at hand. In this case, the functional relationship is given by Equation (3.2). Under the assumption of linearity, the equation suggests that high yield will be obtained at high temperature, low concentration, and with catalyst B.
3. To determine whether all effects are linear or not. This objective was not accomplished in this experiment because a center point trial was not conducted to determine curvature. Therefore, we do not know if the optimum suggested above under the assumption of linear effects is valid or not.

A two-level designed experiment necessarily assumes that the effects are linear, whether they truly are or not. To test this assumption of linearity, at least three levels of each factor will have to be considered, which will make the experiment very large. A compromise approach is to only conduct one additional trial, usually at the center of the experimental region. Under the assumption of linearity, the predicted value at the center will simply be the average of all data. If the actual experimental result is close to this predicted value, the effects could reasonably be assumed to be linear, otherwise not. While the center point identifies curvature, from this single trial it is not possible to tell which factors are responsible for the curvature. Additional trials are necessary and these trials constitute part of the optimization experiment, as we shall see later. In this example, because catalyst is a discrete factor, a trial at the center of the cube cannot be conducted. But a center point in the other two continuous factors should have been conducted to assess if the effects of temperature and concentration are linear. Not conducting such a trial was a mistake.

3.3 SUCCESS FACTORS

If designing and analyzing experiments is so simple, why do all designed experiments not succeed? The success of the experiment has much to do with making correct scientific and engineering judgments and less to do with statistics. With software packages, it is now easy to mechanically execute the job of designing and analyzing experiments. The downside is that people sometimes forget to think! The most important success factors, namely, what to measure, which factors to include in the experiment, what the ranges for these factors should be, and what type of an experiment to structure, are questions that need thinking and knowledge. All available scientific and engineering knowledge should be used while designing the experiment. Some of these key success factors are briefly discussed below.

What to Measure?   The experiment design begins by first selecting the responses we have to measure. Then, we select the control factors because their selection depends upon the selected responses. Here are some desirable properties for responses.

1. Responses should measure the ideal function of the system without being influenced by factors outside the system being optimized. If the function of the system is to break up large particles into smaller particles, then particle size is the correct response to measure. Yield defined as the percent of particles within specification is not the right response because yield depends upon the specification. The desired size range, while it is a customer requirement, is not a direct measure of the physics of the system. In a spray-coating process, the main function is to obtain uniform coating thickness. Therefore, coating thickness is the correct characteristic to measure, and not sags and dripping paint, which are a function of gravity, or whether the result is acceptable or not.

2. The selected responses should form a complete set encompassing product function, side effects, and failure modes. For example, if we measure a quality attribute, we should also measure cost.

3. Responses measured on a continuous scale are preferred. They require smaller sample sizes. If attribute measurements are used, sample sizes can be very large. To reduce sample sizes for attribute responses, worst-case testing could be done.

4. Responses should be easy, inexpensive, and precise to measure. This will make the experiment practical to conduct. As necessary, a measurement system study

should be performed (see Chapter 9) to understand and reduce measurement variability.

5. Responses should be chosen to minimize control factor interactions. Choice of a wrong response can engender control factor interactions, making the system look unnecessarily complex, and lead to large experiments and complicated models. This concept will be discussed in more detail in Chapter 4.

What to Analyze?   Just because we decide to measure (say) pull strength does not necessarily mean that we should analyze pull strength. Statistical data analysis makes two key assumptions that the mean and standard deviation of residuals are independent of each other and that the residuals of the fitted equation are normally distributed. Often these two issues are coupled. For pull strength, as the mean increases, so does the standard deviation. Pull strength also has a positively skewed distribution. In such situations, it is important to transform the observed pull strength values (e.g., by taking the log of pull strength) such that for the transformed data, the mean and standard deviation are independent. The transformed data also has a more normal distribution. Analysis of a properly transformed data often leads to simpler models, with fewer interaction terms. The transformed data are analyzed and the results are reverse transformed and presented in the usual, understandable units. The specific transformation to be used depends upon the relationship between the mean and standard deviation. Some commonly used transformations, known as the Box–Cox transformations (Box et al., 1978), are in Table 3.4 where Y represents the original data and Z represents the transformed data.

Selecting Control Factors   The second step in the classical design of experiments is to select control factors that are likely to influence the selected responses. The following are some guidelines for selecting control factors:

1. All potentially important factors should be included in the experiment. The factors should not be dropped just because we think that the experiment will become unmanageably large. If an important factor is not included in the experiment, the chances of success reduce. Increasing the number of factors in

TABLE 3.4 Transformations

Transformation	Relationship Between σ and μ	Application
$Z = 1/Y$	$\sigma \propto \mu^2$	
$Z = 1/\sqrt{Y}$	$\sigma \propto \mu^{3/2}$	
$Z = \log Y$	$\sigma \propto \mu$	Variance, or when percent error is constant
$Z = \sqrt{Y}$	$\sigma \propto \mu^{1/2}$	Number of defects
$Z = Y$	$\sigma \propto$ constant	
$Z = \text{arcsine}\sqrt{Y}$		Fraction defective

an experiment becomes easier, when it is understood that it does not dramatically increase the size of the experiment. This is explained below under replication.

2. Continuous control factors are preferred. If a control factor is discrete, such as type of material, every effort should be made to convert it to a continuous control factor through proper scientific and engineering understanding based upon material properties.

3. Control factors should be selected to influence both the mean and the variability of response. It is desirable to have many factors that will potentially reduce variability and at least one factor that will allow mean to be changed without affecting variability. This is illustrated by the drug-coated stent example considered below.

4. If multiple control factors have the same function, they will interact with each other, unnecessarily complicating the system. If they are redundant, only one factor from this group may be used, or a joint factor can be created.

Selecting Control Factor Levels   For screening experiments, only two levels for each control factor should be selected unless the factor is known to have a curvilinear effect with a maximum or minimum in the experimental range. If the experiment is intended to identify key factors, the factors should be varied over a wide yet realistic range to magnify effects making their detection easier. Control factor levels should be selected to minimize known control factor interactions as explained at the end of this chapter and in Chapter 4.

Center Points   A center point in continuous factors should be included in a screening experiment to detect curvature. Effects of one or more factors may not be linear if the predicted value at the center departs meaningfully from the experimentally observed value. In this case, the screening experiment is augmented with additional trials to determine the factors causing curvature as explained later in this chapter under optimization experiments.

Replication   Replication means repeating the entire or a portion of the experiment one or more times. Replication increases the number of trials and is expensive, but may be necessary for several reasons. We consider one reason here. If we have to detect an effect of size Δ and the replication standard deviation is σ, then the approximate total number of necessary trials is

$$n_{\text{total}} = \frac{64}{d^2} \quad \text{where} \quad d = \frac{\Delta}{\sigma} \tag{3.3}$$

For example, if we believe that an effect as small as 2 is practically important, then $\Delta = 2$. If σ is estimated to be 1, then $d = 2$ and $n_{\text{total}} = 16$. We will need to conduct a total of 16 trials, that is, a 2^3 factorial will have to be completely repeated once. This total number of trials is essentially independent of the number of factors in the experiment. Increasing the number of factors in the experiment does not significantly increase the number of trials.

There is a difference between replication and duplication. Replication means repeating the trial completely. Duplication means conducting the trial once and taking duplicate measurements of the product produced. Duplication is cheaper, but can we get by with duplicates rather than replicates? The answer depends upon the reasons for replicate variability. If the principal source of variability is measurement error, duplicates are enough. Otherwise, replication is necessary. The main point is that it is the key causes of replicate variability that need to be repeated. Variance components analysis helps identify these key causes and develop a cost-effective sampling scheme as shown in Chapter 9.

Randomization   Randomization means conducting experiments in a random order. The purpose of randomization is to help prevent unknown causes of variation from biasing the estimated effects. An experiment should be run in a random order whenever possible. However, it is not possible to effectively randomize compact experiments such as an unreplicated fractional factorial design because the random order is likely to correlate with some factor or interaction. Also, at times one may have to depart from a random sequence for cost or time reasons. This will be the case if one or more factors are difficult to change. The economic consequences of conducting the trials in a certain sequence may far outweigh the benefits of randomization.

Blocking   Blocking reduces the impact of known causes of variability and thereby improves the accuracy and precision of estimated effects. Suppose a lot of raw material is sufficient to run four trials and we have to conduct an eight-trial 2^3 factorial in three other factors A, B, and C. If the three-factor interaction ABC could be assumed to be small, then trials where this interaction is at the -1 level could be assigned to lot 1 of the raw material and the remaining trials to lot 2 of the raw material. The effect of the raw material lot differences and the effect of the three-factor interaction ABC could not be separately estimated from this experiment, but the experiment will avoid the raw material lot differences from influencing the estimation of other effects.

Conducting Experiments   If it appears that some trials may be impossible to run or may produce very questionable results, they should be run first or the ranges of control factors should be properly adjusted. A decision should be made regarding whether the trials should be conducted in a random order or in an order that is easy to execute. If a trial is run at incorrect settings, often it may only be necessary to note the settings and analyze the modified experiment. However, for highly fractionated experiments and for large errors in factor settings, the trial may have to be rerun at the original settings. All trials of the designed experiment should be conducted. This is particularly the case for a highly fractionated fractional factorial. If some replicated trials are missing, they need not be rerun.

3.4 FRACTIONAL FACTORIAL DESIGNS

The factorial design approach is not practical for large number of factors because the number of trials become prohibitively large. Seven factors will require $2^7 = 128$ trials. On the other hand, the main effects of seven factors can be determined in only eight trials. The minimum number of necessary trials is always one more than the number of main and interaction effects to be determined. This means that we should be able to dramatically reduce the size of the experiment, so long as we are willing to give up on certain interactions. This is what a fractional factorial design does. As the name suggests it is a fraction of the full factorial design, usually a very small fraction. It allows a large number of factors to be investigated in a compact experiment. This compactness does not come free, some or all interactions cannot be estimated. This inability of a fractional factorial design to independently estimate all main and interaction effects is known as *confounding* and forms the main new concept to be learnt.

Consider the case where we have four control factors. The factorial design will require 16 trials. Four factors in eight trials is a half-fraction. Since there are four factors and only eight trials, the design is called a 2^{4-1} fractional factorial. Referring to Table 3.5, we know how to deal with three factors in eight trials as a full factorial

TABLE 3.5 Confounding in Fractional Factorials

	2^3 Factorial							
Trial	A	B	C	AB	AC	BC	ABC	Y
1	−	−	−	+	+	+	−	
2	+	−	−	−	−	+	+	
3	−	+	−	−	+	−	+	
4	+	+	−	+	−	−	−	
5	−	−	+	+	−	−	+	
6	+	−	+	−	+	−	−	
7	−	+	+	−	−	+	−	
8	+	+	+	+	+	+	+	

	2^{4-1} Fractional Factorial				
Trail	A	B	C	D	Y
1	−	−	−	−	
2	+	−	−	+	
3	−	+	−	+	
4	+	+	−	−	
5	−	−	+	+	
6	+	−	+	−	
7	−	+	+	−	
8	+	+	+	+	

experiment. So the only real question is: What should be the sign structure for factor D in the 2^{4-1} fractional factorial? In answering this question, we must preserve the property of balance. There are only seven balanced sign structures for an eight-trial experiment and these are shown for the 2^3 factorial in Table 3.5. Three of these sign structures are already used in the 2^{4-1} fractional factorial. So the only choices for the sign structure for factor D are the once under columns titled AB, AC, BC, and ABC. Let us use the ABC sign structure for factor D. The designed 2^{4-1} fractional factorial experiment is shown in Table 3.5 and it is a balanced experiment.

As explained before, the effect of a factor is determined by attaching the sign structure to the data and dividing it by half the number of trials. Since the sign structures for D and ABC are the same, the main effect of D and the effect of ABC interaction will be estimated to be the same. The effects are completely correlated and cannot be separately estimated from this experiment. This is what is meant by confusion in a fractional factorial, and this confusion is represented in a shorthand notation by D = ABC. The story of confusion is a three-step story: determining the confounding structure, deconfounding the results, and managing the confounding structure.

Determining the Confounding Structure   Four factors can have 15 main and interaction effects: 4 main effects, 6 two-factor interactions, 4 three-factor interactions, and 1 four-factor interaction. Hence, we need the full factorial experiment with 16 trials to find them all. From the eight-trial fractional factorial, we can find only seven effects. Just because we can find only 7 effects does not mean that the 15 possible effects have reduced in number. So, there is bound to be considerable confusion in this experiment.

Table 3.6 shows the confounding structure for the 2^{4-1} fractional factorial experiment. We decided to use the ABC interaction column for factor D, resulting in D = ABC. If we multiply both sides of this equation by column D, we get $D^2 = ABCD$. But D^2 (column D multiplied by itself) is a column of $+1$ and is called identity I. So the identity equation becomes I = ABCD. Multiplying both sides of this identity equation by the effects to be estimated leads to the confounding structure in Table 3.6. For example, multiplying both sides of the identity equation by A leads to $A = A^2BCD$, and A^2 is just a column of $+1$ and can be dropped.

TABLE 3.6   Confounding Structure

Design derivation, D = ABC
Identity, I = ABCD
Confounding structure
A = BCD
B = ACD
C = ABD
D = ABC
AB = CD
AC = BD
AD = BC

The main effects are confused with the three-factor interactions and the two-factor interactions are confused in pairs. The four-factor interaction is confused with identity, that is, it never changed in the experiment and cannot be estimated. Software packages easily provide the confounding structure for fractional factorial designs.

Deconfounding the Results   Effects can be estimated by using the usual procedures. For example, the effect AB = CD is estimated by constructing a column of the AB interaction, attaching the sign structure to the data, and dividing by 4. The Pareto principle applies and only a few effects will be important. Suppose it turns out that the effects A = BCD, B = ACD, and AB = CD are important. Deconfounding means finding out which of the confounded factors and interactions are responsible for the observed effects.

There are three ways to resolve the confusion—use rules of thumb, use technical knowledge, and if these fail, conduct additional trials.

The first approach to deconfounding is to use two rules of thumb. The first rule states that the three-factor and higher-order interactions are usually small. This means that the main effects of A and B are important and not the BCD and the ACD interactions. The second rule states that the important two-factor interaction is between factors whose main effects are important. This means that it is the AB interaction that is important and not the CD interaction. These rules of thumb often resolve the confounding structure as in this case where the important effects are likely to be the main effects of A and B, and the AB interaction. Sometimes these rules of thumb fail to completely resolve the confusion. If AC = BD had turned out to be important, then the second rule of thumb will fail to resolve this confusion. The fact that both A and B have turned out to have large effects will make it difficult to decide whether it is the AC or the BD interaction that is important.

The second approach is to use our own technical understanding of the subject. To resolve AC = BD, the AC and BD interactions can be graphically represented as shown in Figure 3.7 so that their meaning is easier to interpret. The interaction that is likely to be true can be chosen on the basis of a technical understanding of the subject matter.

If the above two approaches fail to resolve the confusion, then additional experiments can be designed to deconfound the results experimentally. Presently, a single additional trial will experimentally resolve the confusion between the AC and the BD interactions.

Managing Confusion   In the experiment, had we chosen the AB interaction structure for factor D then the identity would have been I = ABD (resolution 3 design) and the confounding structure would have been different. We do have some control over the confounding structure in an experiment. By managing confusion, we mean to achieve a confounding structure that will be easier to deconfuse later.

Let us suppose that we have to find the main effects of eight factors A, B, C, ..., H, and based upon a prior technical assessment we think that the interactions AB, AC, and BC may be important. Then, the 16-trial fractional factorial should be designed such that the specified main effects and interactions are not confused with each other. The deconfounding will be easier as opposed to the case where the confounding structure includes H = AB.

Design Resolution   The higher the design resolution, the easier it usually is to resolve confusion. The above fractional factorial design is said to be a resolution 4 design because the smallest word in the identity is a four-letter word. The result is that the main effects are confused with the three-factor interactions and the two-factor interactions are confused with each other. Such a confounding structure is easier to resolve using the rules of thumb. A resolution 3 experiment can be much more compact, and therefore, less costly to conduct. For example, seven factors in eight trials is a resolution 3 experiment. However, for resolution 3 experiments, the identity is a three-letter word, that is, the main effects are confused with the two-factor interactions. This confounding structure is more difficult to resolve. Clearly, if the total number of trials is the same, resolution 4 design is better than the resolution 3 design. Eight factors in 16 trials is a resolution 4 design. Nine factors in 16 trials is a resolution 3 design. Nine factors in 32 trials is a resolution 4 design. How many factors and trials to use, and what risks and costs to incur, is a practical decision to be made based upon technical and business judgments.

3.5   PLACKETT–BURMAN DESIGNS

Plackett–Burman designs, invented in 1946, are another class of useful screening designs. While the number of trials n in a two-level factorial or fractional factorial design is 2^k where k is an integer, the number of trials in the Plackett–Burman design is $4k$. For most industrial experiments, Plackett–Burman designs with 12, 20, 24, 28, and 36 trials are a useful addition to the array of available fractional factorial designs. For example, to investigate the main effects of 9 factors, the smallest number of necessary trials is 10. The nearest available fractional factorial design has 16 trials. Instead, a 12-trial Plackett–Burman design may be used saving 25 percent of the experimental effort.

Table 3.7 shows a 12-trial Plackett–Burman design. This is a balanced design. The main effects of up to 11 factors can be estimated using this experiment. However, it is

TABLE 3.7   Twelve-Trial Plackett–Burman Design

Trial	1	2	3	4	5	6	7	8	9	10	11
1	+	+	−	+	+	+	−	−	−	+	−
2	+	−	+	+	+	−	−	−	+	−	+
3	−	+	+	+	−	−	−	+	−	+	+
4	+	+	+	−	−	−	+	−	+	+	−
5	+	+	−	−	−	+	−	+	+	−	+
6	+	−	−	−	+	−	+	+	−	+	+
7	−	−	−	+	−	+	+	−	+	+	+
8	−	−	+	−	+	+	−	+	+	+	−
9	−	+	−	+	+	−	+	+	+	−	−
10	+	−	+	+	−	+	+	+	−	−	−
11	−	+	+	−	+	+	+	−	−	−	+
12	−	−	−	−	−	−	−	−	−	−	−

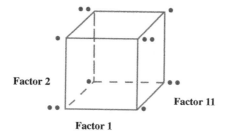

Factor 2

Factor 11

Factor 1

FIGURE 3.8   Plackett–Burman design as a full factorial in any three factors.

not possible to determine any interactions. The confounding structure for this design is very complicated with bits and pieces of several interactions confused with the main effects in each column.

This inability to estimate an interaction has resulted is less use of these designs in the past. However, in the 1990s, an interesting property of Plackett–Burman designs was discovered (see Box and Tyssedal (1996)). It was shown that these Plackett–Burman designs can be reanalyzed as full factorial experiments in any three factors. This is shown in Figure 3.8 graphically for factors 1, 2, and 11. Thus, the 12-trial Plackett–Burman design is a full factorial in these three factors with four replicates, which themselves form a half fraction.

This allows us to reanalyze the experiment as follows. Suppose the original experiment involved 11 factors in 12 trials. Let us further suppose that up to three factor effects were large. Then, the experiment can be reanalyzed as a factorial in these three factors to assess the significance of interactions between the three factors.

It should be noted that the fractional factorial designs also have a similar property. Any fractional factorial design of resolution R can be reanalyzed as a full factorial in any $(R - 1)$ factors.

3.6   APPLICATIONS

Let us now consider some applications of fractional factorial experiments. The first example illustrates how the methodology can be used to develop robust processes. The second example illustrates a nontraditional application to marketing.

Drug-Coated Stent Example   The purpose of this experiment was to reduce the variability of release rate of the drug-coated stent while maintaining the average release rate on target. Both process and formulation factors were included in the experiment. The four control factors, their ranges, and the observed values of average release rate and the CV of release rate are shown in Table 3.8. A half-factorial design was used and each trial was repeated six times to obtain the reported average and CV of release rate. The last trial is the center point, which was the current operating condition for the process.

TABLE 3.8   Drug-Coated Stent Example

Trial	Stent–Nozzle Distance	Atomization Pressure	Forward Speed	Formulation Factor	Mean	CV
1	10	5	8	60	87	3.9
2	10	5	16	80	42	11.9
3	10	20	8	80	15	9.4
4	10	20	16	60	25	11.3
5	20	5	8	80	85	1.3
6	20	5	16	60	74	8.9
7	20	20	8	60	37	5.5
8	20	20	16	80	20	9.3
9	15	12.5	12	70	47	7.9

The experiment was analyzed to determine the important effects using the methods described previously. The analysis was conducted without including the center point trial. The mean release rate was shown to be influenced by all the four main effects and also by the interaction between atomization pressure and forward speed. The same was true regarding the CV of release rate except that the formulation factor had no effect on CV. The major implication is that formulation factor can be used to change the mean release rate without affecting the CV of release rate.

In this experiment, the interaction between atomization pressure and forward speed is confounded with the interaction between stent–nozzle distance and formulation factor. This confounding was resolved by noting that atomization pressure and forward speed have the largest main effects while formulation factor has no effect on CV.

Based upon this analysis, the following two equations were derived. They are written in coded units where every factor ranges from -1 to $+1$.

$$\text{Mean} = 48.1 + 5.9 \text{ stent-nozzle distance} - 23.9 \text{ atomization pressure}$$

$$-7.9 \text{ forward speed} - 7.6 \text{ formulation factor} + 6.1 \text{ atomization pressure}$$

$$\times \text{ forward speed} + \text{residual} \qquad (3.4)$$

$$\text{CV} = 7.7 - 1.4 \text{ stent-nozzle distance} + 1.2 \text{ atomization pressure}$$

$$+ 2.7 \text{ forward speed} - 1.2 \text{ atomization pressure}$$

$$\times \text{ forward speed} + \text{residual} \qquad (3.5)$$

Under the assumption of linearity, the predicted mean and CV at the center point are 48.1 and 7.7, respectively. This can be easily seen from the two equations because at the center point, the coded value of all control factors is zero. From trial 9 in Table 3.8, the experimentally observed center point values are 47 and 7.9, respectively. Since the predicted values are very close to the observed values, the assumption of linearity is justified.

The objective was to achieve a target release rate of 50 with the smallest possible CV. At the center point, which was the current operating condition for the process before this experiment, the mean is close to the target, but the CV is large. We have to solve the above two equations to find the levels of the four control factors where the desired mean and a smaller CV are obtained. From the equations, it is clear that increasing the formulation factor reduces the mean without any effect on CV. Increasing the stent–nozzle distance reduces CV, but increases the mean. The effects of the other two factors are more difficult to see because they interact. Their effect is best seen by plotting the response surface contours plots of the above two equations as shown in Figure 3.9(a) and (b). For the contour plots in Figure 3.9, the other two factors were held constant at the stent–nozzle distance = 20 (to minimize CV) and formulation factor = 80 (to reduce mean). For a mean release rate of 50, minimum CV = 4 is obtained at forward speed = 8 and an atomization pressure = 13.5. Within the experimental zone, these levels of the four factors are the best and represent a 50 percent reduction in CV compared with the current operating condition. Further experiments could be done outside this experimental zone, for example, at lower

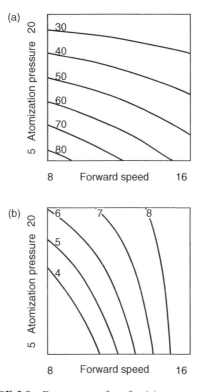

FIGURE 3.9   Response surface for (a) mean and (b) CV.

values of forward speed to reduce CV further. In doing so, other practical issues such as productivity, cost, and impact on other responses of interest will have to be considered.

Marketing Application   Although most of the design of experiment applications are in the areas of science and engineering, the subject can be fruitfully applied to other areas as well. In the food industry, concept tests are conducted to determine the appeal of certain food concepts to consumers. This is often done by writing the contemplated food concepts in an advertising copy format, and then, getting the reaction of a selected consumer panel to a large number of concepts. From one concept to the next, many product attributes such as price, nutrition, and convenience are simultaneously changed. Because these changes in the contemplated product attributes are not made in a balanced fashion, it is often not possible to make unambiguous statements regarding the relative importance of the product attributes. Many concept tests fail to provide clear direction to R&D regarding which product attributes are the most important to the consumer. This situation could be changed by first designing a fractional factorial using the contemplated product attributes as control factors, and then, writing a concept for each designed trial of the fractional factorial. The mean and standard deviation of consumer responses could then be analyzed to determine the relative importance of each product attribute. Interactions between the product attributes can also be determined to identify strong synergistic or antisynergistic attributes. The robustness of the product attributes can be assessed as well.

Let us now consider an application involving getting customers to respond to a national campaign to sell a particular software product. We have to maximize the response rate. Small improvements in response rate can have large effects on profitability. Four variables are considered: price, message, promotion, and free upgrades. The levels of each variable are the following: price (320, 340, 360, and 380), message (user-friendly, powerful), promotion (none, 30-day free trial), upgrades (none offered free, free upgrades for a year). The total number of combinations is 32, but an eight-trial fractional factorial can be designed and is shown in Table 3.9. The control factor levels are denoted by 1, 2, and so forth, and the measured response

TABLE 3.9   Marketing Experiment

Trial	Price	Message	Promotion	Upgrades	p	$\log\left(\dfrac{p}{(1-p)}\right)$	Predicted p
1	1	1	1	1	0.10	-0.95	0.12
2	1	2	2	2	0.38	-0.21	0.38
3	2	1	1	2	0.12	-0.87	0.13
4	2	2	2	1	0.08	-1.06	0.08
5	3	1	2	1	0.04	-1.38	0.03
6	3	2	1	2	0.07	-1.12	0.06
7	4	1	2	2	0.03	-1.51	0.04
8	4	2	1	1	0.01	-2.00	0.01

rate p is the number of responses received divided by the number of solicitations sent, expressed as a fraction.

If we analyze p as a response, then it is possible that the resultant equation may predict negative response rates or response rates greater than 1.0 (100 percent) for certain combinations of input factors. Since p is a fraction defective type response it needs to be transformed before analysis. One appropriate transformation to use is $\log(p/(1-p))$. The arcsine transformation in Table 3.4 may also be used. The standard analysis of this designed experiment shows the effect of message to be statistically and practically insignificant, that is, there is no difference between the user friendly and the powerful message. This does not mean that no message is necessary, simply that either message is equally effective. To determine whether message is necessary at all, the level no message will have to be tested, or could have been tested by making message a three-level factor. The fitted equation is

$$\text{Predicted } \log\left(\frac{p}{(1-p)}\right) = -1.14 - 0.57 \text{ price} + 0.097 \text{ promotion} + 0.21 \text{ upgrade}$$

$$(3.6)$$

Equation (3.6) can be used to predict the response rate for any combination of price, promotion, and upgrade. The predicted values of p, obtained by reverse transformation, are shown in the last column of Table 3.9. The equation shows that providing free upgrades for a year is more important than a 30-day free trial and that the response rate is very sensitive to price. This is targeted information to achieve high response rates. In another situation, the results may have been completely different.

Further points should be made with respect to the design of this experiment and the determination of the total number of respondents in the study. It was not necessary to have four levels of price; two levels would have been enough. If for some reason three levels of each control factor were desired, a three-level fractional factorial design could have been constructed in nine trials. This design will not identify interactions between factors. If interactions between certain factors were suspected, the experiment could have been appropriately designed to find the interactions. The necessary total number of respondents can be approximately obtained from Equation (3.3). Before conducting the experiment, suppose the average response rate was expected to be 0.1 (10 percent). Then the standard deviation of a single observation (based upon binomial distribution) is $\sqrt{(0.1)(0.9)} = 0.3$. If a change in response rate as small as 0.05 (5 percent) is to be detected, then $\sigma = 0.3$ and $\Delta = 0.05$, that is, $d = 1/6$. This gives the total number of respondents as $64/d^2 = 2300$. Approximately 300 respondents per trial could be used. The respondents need to be a random sample from the target population.

3.7 OPTIMIZATION DESIGNS

Design of experiments is a two-step process—screening followed by optimization. If the center point shows a lack of curvature, then the screening experiment itself becomes the optimization experiment. This was the case with the drug-coated stent example. In that experiment, had the center point experimental result deviated markedly from the average of the other eight trials, it would have indicated a curvilinear effect due to one or more control factors. Additional optimization trials would have been necessary to identify the factors causing curvature.

The design and analysis of optimization experiments differs from that of the two-level screening experiments in two ways. One change is that each factor needs to have at least three levels in order to determine curvature effects. One experimental strategy is to use three-level factorial designs. But this strategy becomes very expensive, because for three factors the number of trials will be $3 \times 3 \times 3 = 27$, for four factors the number of trials will be 81, and so on. There are three types of commonly used optimization designs—central composite designs, Box–Behnken designs, and three-level fractional factorial designs. The commonly used three-level fractional factorial designs do not permit the determination of interactions and these designs should be used when the main and quadratic effects are expected to be much larger than the interaction effects. The central composite design is preferred to the Box–Behnken design because it allows a previous screening experiment to be used as a part of the optimization experiment. Central composite design is often the design of choice. The second change is that the fitted model needs to include quadratic terms in addition to the main effect and interaction terms. The design and analysis of optimization experiments is comprehensively discussed in Box and Draper (1987). We briefly consider the practical central composite design.

A three-factor central composite design is shown in Figure 3.10. A central composite design consists of a factorial (or a fractional factorial) with a center point, to which star points are added as shown in Figure 3.10. The star points are located on perpendiculars drawn from the center on to each face of the cube. They are generally located either on the face of the cube or outside the cube. The combined experiment is known as the central composite design. For three factors, this design has 15 trials. Each factor has five levels, which will reduce to three levels if the star points are located on the face of the cube. Among the many advantages of a central composite design is the fact that the factorial design and the center point may already be done as a screening experiment. If the center point indicates curvature, the screening experiment simply needs to be augmented by adding the six star points. We do not need 15 new trials, we only need 6.

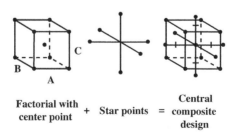

Factorial with center point + Star points = Central composite design

FIGURE 3.10   Central composite design.

Let us now consider a simple application of an optimization experiment with only two factors, time and temperature with yield as response. Experiments involving time and temperature as control factors are very common in industry. The designed experiment, using star points located at the factorial boundary, is shown in Table 3.10 along with yield measured in pounds.

Trials 1–4, which form the factorial experiment, and trial 9, which is the center point, had been previously conducted as a screening experiment. When curvature was detected, the remaining four trials were added to determine which factor was responsible for the curvature.

The usual analysis of the entire experiment identified that the curvature was due to time and led to the following equation in coded units:

$$\text{Yield} = 299 - 1.8 \text{ time} + 5.7 \text{ temperature} - 109.8 \text{ time}$$

$$\times \text{ temperature} - 112.8 \text{ time}^2 + \text{residual}$$

Clearly, the interaction between time and temperature and the quadratic effect of time dominate the equation. Even though this is a simple quadratic equation, it is not easy to visualize it. The implications of this equation are best seen by plotting the response surface contours (Figure 3.11). The response surface contours have a saddle shape, with two maxima, one at high temperature and lower time and the other at low temperature and high time. The maximum yield of around 340 pounds is obtained at temperature = 360 and time = 10. This combination of time and temperature gives higher yield than the other maximum and is also less costly due to the shorter time.

Often we have to optimize multiple responses simultaneously. For example, we may have to maximize yield and minimize cost. When there are multiple responses, each response is individually analyzed as explained here. To conduct multiresponse optimization, the individual response surface contours can be overlaid on top of each other to identify the experimental region where all objectives are satisfied. Software packages make this simpler using search and other algorithms.

The complicated response surface for yield resulted from the large negative interaction between time and temperature. This large negative interaction should

TABLE 3.10   An Optimization Experiment

Trial	Time	Temperature	Yield
1	5	180	72
2	5	360	306
3	30	180	288
4	30	360	83
5	5	270	185
6	30	270	181
7	17.5	180	295
8	17.5	360	300
9	17.5	270	301

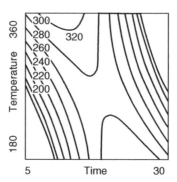

FIGURE 3.11   Response surface contours for yield.

have been anticipated while designing the experiment. In experiments involving time and temperature, very often it is the total energy that matters. Right level of energy is necessary to get good yield. This right level of energy could be obtained with high temperature and low time combination or low temperature and high time combination. If this issue was considered with sufficient theoretical knowledge, perhaps an experiment could have been designed in a single factor called energy. Sometimes, the rate of input of energy, governed by temperature, also matters. With this knowledge, the experiment could have been designed to avoid the interaction by making the levels of time a function of temperature. These are examples of the use of technical knowledge to eliminate known control factor interactions. These considerations are particularly important in compact experiments involving large number of factors because if many control factors start interacting with each other, a large experiment will be necessary to find these interactions and the resultant complicated equation will be very difficult to understand.

Mixture Designs   In certain situations, as in the case of formulation development experiments, the formulation factors (control factors) add up to 100 percent. These situations are known as mixtures. If a formulation consists of three components A, B, and C, then A + B + C = 100 percent. This mixture constraint means that the levels of A, B, and C cannot be independently selected, and therefore, screening and optimization experiments discussed previously cannot be conducted in A, B, and C. This situation was encountered in Chapter 1 where the portfolio consisted of three asset classes: large company stocks, international stocks, and corporate bonds. Mixture designs have been developed to treat these situations (see Cornell (2002)).

The simplest way to deal with mixtures is to try to get out of the constraint. This can usually be done by taking one or more ratios. For example, if the mixture consists of high molecular weight polymer, low molecular weight polymer, and a solvent, then the problem could be recast in terms of two ratios: high molecular weight to low molecular weight ratio and total polymer to solvent ratio. The ratios can be independently set making usual screening and optimization experiments feasible. In many of these experiments, ratios are far more meaningful than the individual proportions. Another example is dough consisting of flour, water, oil, salt, and several other ingredients that add up to 100 percent. This mixture constraint can be removed by considering flour/ water ratio instead of flour and water separately. The ratio is far more meaningful than the individual proportions of flour and water from the viewpoint of dough viscosity. A second approach to remove the mixture constraint is to eliminate an inert factor, if one exists, from consideration. The remaining factors do not have to sum up to 100 percent and changes in their sum are compensated for by changes in the level of the inert factor. These ways of removing the mixture constraint have the added benefit of being able to experiment with mixture and nonmixture factors simultaneously in a compact experiment.

3.8 QUESTIONS TO ASK

Design of experiments is a very important tool for R&D and manufacturing and every effort should be made to implement the methodology. Is your company using design of experiments effectively? To guide and help implement design of experiments, here are some practical questions to ask:

1. Is the problem well thought through from a scientific and engineering viewpoint?
2. What are the key responses of interest? Are all responses considered? Which are the most important?
3. If the selected responses are discrete, has an effort been made to convert them to continuous responses?
4. Are all potentially important control factors included?
5. Are the selected ranges for control factors wide enough?
6. Is a center point(s) included in the experiment? If not, why?
7. Which factors might interact with each other? Can these interactions be reduced by proper consideration of responses and control factors ranges?
8. Is the experiment designed to ensure that the selected main effects and likely interactions are not confused with each other?
9. What is the size of the smallest effect of practical importance? Are the total number of trials determined based upon this effect size and a prior estimate of replication variance?
10. If this is an optimization experiment, was a previous screening experiment done and did the center point indicate curvature? If a prior screening experiment was not done, why not?
11. Is there a protocol describing the trials and the order in which they will be run? Does the order of experimentation consider the ease of conducting experiments?
12. If this is a mixture experiment, was an attempt made to convert the problem to a nonmixture situation, for example, by using ratios?

13. Are all key analyses done—identifying statistically and practically important effects, understanding and resolving interactions, building adequate equations, estimating curvature, plotting response surface contours, and conducting multiresponse optimization?

14. Were the objectives of the experiment met? What are the next steps? For example, was an input factor missed in the experiment? Were the selected ranges correct? Is there curvature requiring optimization experiments?

Further Reading   Prof. George Box has made many contributions in the field of design of experiments and has been at the forefront of popularizing the use of classical design of experiments. A large portion of this chapter is based upon his teachings and writings. I highly recommend his books: Box et al. (1978) and Box and Draper (1987). The book by Cornell (2002) deals with mixture designs in considerable detail.

CHAPTER 4

WHAT IS THE KEY TO DESIGNING ROBUST PRODUCTS AND PROCESSES?

The robust design method adds two important dimensions to the classical approach to design of experiments discussed in the previous chapter. The first is an explicit consideration of noise factors that cause variability and ways to design products and processes to counteract the effects of these noise factors. The second are ways to improve product transition from research to customer usage such that a design that is optimal at the bench scale is also optimal at the manufacturing scale and during customer usage. This chapter explains the basic principle of achieving robustness and how robustness experiments should be designed and analyzed. It also explains ways to improve scale-up by reducing control factor interactions. These two dimensions have major implications toward how R&D should be conducted.

The robust design method, developed by Dr. Taguchi, focuses on reducing product variability as experienced by the customers. Factors that cause variability are known as noise factors. One usual approach to such problems is to identify the causes of variability and then to either attempt to remove these causes or constrain them with tight specifications. This approach can be very expensive. The objective of the robust design method is to design the product and the process such that the effect of noise factors on responses of interest is minimized, making the product and process designs robust by permitting wider tolerances on noise factors.

Classical design of experiments has often been concerned with expressing the mean response as a function of control factors. Noise factors were not explicitly considered. Often noise has been treated as a nuisance, to be "blocked" or "randomized" to prevent

Industrial Statistics: Practical Methods and Guidance for Improved Performance   By Anand M. Joglekar
Copyright © 2010 John Wiley & Sons, Inc.

it from vitiating our understanding of the relationship between control factors and responses, or to be tightly controlled to reduce variability. Classical design of experiments can also be used to develop robust products and processes by reducing the effect of noise factors, as was demonstrated by the coated stent example in the previous chapter. However, in that example, the effect of noise was assessed through replication, not through explicit identification of noise factors. The explicit introduction of noise factors changes the way experiments need to be designed and analyzed.

The second important objective of the robust design method is to ensure that the design that is found to be the best in a laboratory continues to be the best during manufacturing and customer usage. This is achieved by reducing control factor interactions, that is, by achieving additivity. Additivity means absence of control factor interactions. If control factors have large interactions, particularly antisynergistic interactions, then it is more likely that the control factors will interact with the additional factors that come into play during scale-up and customer usage. This will mean that the best design on the bench scale will no longer be the best design during manufacturing and customer usage. The classical design of experiments approach has typically not been concerned with reducing control factor interactions to the fullest possible extent.

The robust design approach has many similarities with classical design of experiments. In particular, the use of fractional factorial designs, calculation of effects, and interpretation of results continue to be similarly done.

The focus of this chapter is on the two important new objectives of the robust design method:

1. How to reduce variability through an explicit consideration of noise factors and how the incorporation of noise factors changes the design and analysis of the experiment.
2. How to achieve additivity and why this objective requires greater care in selecting responses, control factors, and their levels.

In accomplishing these objectives, the robust design method demands a more extensive use of scientific and engineering knowledge.

4.1 THE KEY TO ROBUSTNESS

The drug-coated stent example in the previous chapter dealt with optimizing the formulation and process factors to achieve on-target release rate with minimum variability. The reasons for the variability of release rate were unknown. Therefore, replication was used as a tool to determine variability under a variety of control factor combinations. Many replicates were necessary to properly observe the impact of the unknown sources of variability. The mean and CV of release rate were analyzed to obtain optimum robust formulation and processing conditions.

What if the factors causing variability of output are known? How can we achieve robustness against these noise factors? What is the key to robustness?

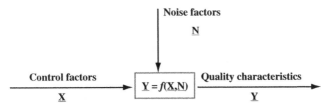

FIGURE 4.1   Designing a robust product.

Problem Formulation   Consider a product such as a patch to transfer drug through skin. We want to achieve a constant rate of drug transfer per hour with the smallest variability. The rate of drug transfer is affected by the properties of the skin—how thick it is, is it oily, the number of pores per unit area, and so on. The skin varies from person to person. Skin is an important factor influencing the rate of drug transfer, yet it is a factor that is not in the hands of the patch manufacturer to control. Such factors are called noise factors, as opposed to patch design factors such as the percentages of formulation components, patch thickness, patch area, and many others that can be controlled by the manufacturer and are therefore known as control factors. By a robust patch we mean a patch that is so designed, namely, the levels of control factors are so chosen that the effect of skin (and other noise factors) on the rate of drug transfer is reduced. If such a robust patch cannot be designed, then the drug transfer rate will vary widely from person to person degrading the functionality of the patch, or the market may have to be segmented by skin type increasing the cost of the patch.

The formulation of the robust product and process design problem is shown in Figure 4.1. The problem is addressed as follows:

1. We begin by defining the key responses or outputs of interest. These responses usually have one of the following objective functions: we want to minimize the output (smaller the better), we want the output to be on target (nominal the best), or we want to maximize the output (larger the better). Impurities are examples of smaller the better characteristics. The rate of drug transfer through skin is an example of nominal the best characteristic. Pull strength is an example of larger the better characteristic.

2. The next step is to identify noise factors that have large effects on the response. Noise factors are classified as follows:
    *   *External Noise Factors:* These are factors that the manufacturer either cannot control or wishes not to control. Customer usage factors such as the skin for drug transfer using a patch and bake temperature and bake time for a frozen pizza are examples of factors that the producer cannot control and have a large effect on product function. Ambient temperature and humidity during manufacturing and raw material variability are examples of process factors that the producer can control but wishes not to for cost reasons and therefore are considered to be process noise factors.

- *Degradation:* Many products degrade over time because of deterioration of the product components due to factors such as ambient conditions, chemical reactions, and mechanical wear. These factors causing product degradation are noise factors. Tool wear is an example of a process degradation factor.

- *Manufacturing Variability:* This is the product-to-product variability at the time of manufacture. Such variation is caused by the variability of key product and process control factors and also by external process noise factors such as ambient conditions during manufacturing. The target values of these key product and process factors are control factors and the variability from these target values is noise. For example, the percentage of a certain ingredient may vary from batch to batch. The target percentage is a control factor and the batch-to-batch variability from target is a noise factor. The target value for a resistor may be 100 ohms (the target value is a control factor) but the individual resistors may vary over a range from (say) 98 to 102 ohms. This variability is a noise factor. The same factor is simultaneously a control and a noise factor. This is also referred to as an *internal noise factor.*

  It should be noted that replicated trials attempt to capture the effects of both the internal and external noise factors. However, these effects are captured through random drawings from the noise space, and as a result, a large number of replicates are necessary to estimate the variance due to noise. On the other hand, when noise factor levels are deliberately varied over a large range, a very small sample size (often equal to one for each noise condition) is sufficient to understand the effect of noise. Another way to express this is that by explicitly experimenting with noise factors, a variance estimation problem is converted into a problem of estimating mean resulting in a dramatic drop in the necessary sample size.

3. Once the responses and noise factors are identified we wish to design the product and the manufacturing process such that the desired values of the responses are obtained while *simultaneously minimizing the effect of the noise factors on the outputs of interest.* Control factors are selected with these two objectives in mind. We want to select control factors that are likely to have a large effect on the responses of interest, and also are likely to counteract the effects of noise factors.

Countermeasures to Noise   Two fundamentally different measures can be taken to counteract the effects of noise factors. One approach is to control noise factors themselves, namely, if ambient temperatures and humidity cause variability, air condition the factory. If large variability in resistors causes variability in response, buy resistors to a tighter specification. This is the idea of "identify the cause of the problem and get rid of the cause." This is tolerance design (see Chapter 5) and it can be an expensive approach.

The second approach is to design the product and the process in such a way that the effects of noise factors are reduced, namely, to design the manufacturing process in such a way that ambient conditions do not influence the response, or to design the electronic circuit in such a way to accept large variability in resistors without influencing the output of interest. This is the idea of "identify the cause of the problem and get rid of the *effect of the cause*." This is the robust design approach. It is an economical way to reduce variability compared to the idea of tightening specifications on noise factors.

Key to Robustness   Robustness means minimizing the effects of noise factors without eliminating the noise factors. This is achieved by the proper selection of the levels of control factors. The key to robustness is the interaction between a control factor and a noise factor. The effect of external noise factors can be reduced if there are interactions between control factors and external noise factors such as skin. The effects of product degradation factors, such as ambient temperature, humidity, light, dust, and mechanical wear, can be reduced if there are interactions between control factors and product degradation factors. One reason for manufacturing variability is because key product design factors (components and ingredients) change from one product to the other. So, the control and noise factors are the same in this case. Reducing the effects of changes in control factors requires interactions of control factors with themselves, namely, the quadratic effects of control factors and the interaction effects between control factors. The main point is that if there are no interactions between control and noise factors, robustness cannot be achieved. This may be seen as follows (Taguchi, 1987).

Let us suppose that the performance of a product is influenced by ambient temperature over which the manufacturer has no control. The manufacturer would like the product to work well over a wide range of ambient temperatures. Ambient temperature is an external noise factor. In Figure 4.2, H_1 and H_3 represent the range of ambient temperature that the product is likely to see in practice. H_2 is the nominal ambient temperature. Let A be a *continuous* control factor (design factor) with A_1 and A_2 being the low and the high levels, respectively. Since A is a continuous control factor, the product could be designed using any desired level of A, not just the levels A_1 and A_2.

Figure 4.2(a)–(c) shows some possible results that may be obtained by evaluating two levels of control factor A at three levels of ambient temperature H. To achieve

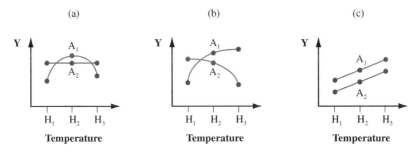

FIGURE 4.2   Robustness against external noise factor, ambient temperature (*AH* interaction).

robustness against temperature, we need to find the level of A such that the effect of noise factor H on the response Y reduces.

From Figure 4.2(a), if we choose A_2 as the level of the control factor A, the effect of ambient temperature disappears. Had we chosen A_1 as the level, the response would have varied in a quadratic fashion as shown. From Figure 4.2(b), the effect of temperature is positive at level A_1 and negative at level A_2. Neither A_1 nor A_2 eliminates the effect of temperature. However, since control factor A is a continuous control factor, we can choose a level of A that is in between A_1 and A_2. While the effect of temperature may not disappear completely, it will reduce substantially. If the situation turns out to be as in Figure 4.2(c), it is not possible to achieve robustness against temperature by using control factor A. The effects at both levels of A are equal, the curves are parallel, and no value of A is likely to reduce the effect of temperature in this case. Control factor A can be used to achieve robustness in Figure 4.2(a) and (b), but not in Figure 4.2(c). Note that control factor A and noise factor H interact in Figure 4.2(a) and (b), but not in Figure 4.2(c). What is necessary to achieve robustness is an interaction between a control factor and a noise factor. The situation with product degradation factors is very similar to the above because these factors are also external noise factors.

Now let us consider the case of manufacturing variability caused by variation of product control factors such as components, ingredients, and so on. Let us suppose that A is a transistor and B is a resistor in a voltage converter. We want the output of the voltage converter to be a constant 110 volts. The effects of A and B on the output voltage are shown in Figure 4.3. Suppose that A has a wide tolerance of ± 20 percent to keep costs down. $A_1 B_1$ is one combination that gives 110 volts output. However, since A varies considerably from one product to the next and the slope of the curve is large at A_1, the output voltage will have large product-to-product variation. Another possibility is to use A_2 as the level for the transistor. In this case, the output voltage will be larger than desired, but this can be compensated for by using B_2 as the level for the resistor. Thus, $A_2 B_2$ is another combination that gives 110 volts output. In this case, a ± 20 percent change in A_2 will have little effect on the output because of the flatness of the curve at A_2. A is a control factor and also a noise factor. Interaction of A with itself, namely, the quadratic effect of A, is necessary to achieve robustness. Robustness

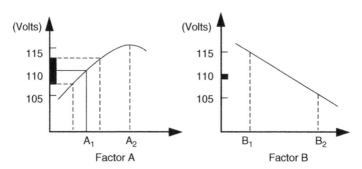

FIGURE 4.3   Robustness against manufacturing variability.

against changes in B cannot be similarly obtained because the effect of B is linear, that is, it does not interact with itself.

4.2 ROBUST DESIGN METHOD

The robust design method is now illustrated by an example. It is assumed that the reader is familiar with design of experiments as explained in the previous chapter. The objective of the experiment was to design a stent coating system to obtain the desired coating thickness with the smallest possible variability. Four external noise factors and eight control factors $(A, B, \ldots, H)$ were identified as relevant factors. The designed experiment and a summary of the observed data are shown in Table 4.1.

How should the robustness experiment be designed? One may think that there are a total of 12 factors, 8 control factors and 4 noise factors. The 12 factors can be evaluated in a 16-trial fractional factorial design. Such an experiment would be a wrong experiment because of the following reason. Robustness can only be achieved if there are interactions between control factors and noise factors. Between the 8 control factors and the 4 noise factors, there are a total of 32 two-factor control-by-noise interactions. The 12 factors in the 16-trial experiment would be inappropriate because it will not permit us to determine the control-by-noise interactions of interest. The experiment needs to be designed in a way to permit control-by-noise factor interactions to be estimated. While there are other ways of designing robustness experiments, one approach is to design the experiment as two separate designs, one in control factors and the other in noise factors, with each control factor trial evaluated at each

TABLE 4.1   Robustness Experiment

Trial	A	B	C	D	E	F	G	H	N_1	N_2	N_3	$\bar{y}$	s	$20\log(\bar{y}/s)$
1	-1	-1	-1	-1	-1	-1	-1	-1	•	•	•	50	12.2	12.3
2	-1	-1	-1	1	1	1	1	-1	•	•	•	51	13.6	11.5
3	-1	-1	1	-1	1	1	-1	1	•	•	•	40	10.5	11.6
4	-1	-1	1	1	-1	-1	1	1	•	•	•	41	9.8	12.4
5	-1	1	-1	-1	1	-1	1	1	•	•	•	52	31.6	4.3
6	-1	1	-1	1	-1	1	-1	1	•	•	•	53	32.7	4.2
7	-1	1	1	-1	-1	1	1	-1	•	•	•	44	27.8	4.0
8	-1	1	1	1	1	-1	-1	-1	•	•	•	43	24.7	4.8
9	1	-1	-1	-1	-1	1	1	1	•	•	•	57	12.8	13.0
10	1	-1	-1	1	1	-1	-1	1	•	•	•	58	11.1	14.4
11	1	-1	1	-1	1	-1	1	-1	•	•	•	47	10.6	12.9
12	1	-1	1	1	-1	1	-1	-1	•	•	•	48	8.4	15.1
13	1	1	-1	-1	1	1	-1	-1	•	•	•	57	18.7	9.7
14	1	1	-1	1	-1	-1	1	-1	•	•	•	59	16.4	11.1
15	1	1	1	-1	-1	-1	-1	1	•	•	•	48	17.1	9.0
16	1	1	1	1	1	1	1	1	•	•	•	49	16.6	9.4

combination of noise factors. Such an arrangement permits all control-by-noise interactions to be determined.

Control Factor Design The control factor design shown in Table 4.1 is a resolution 4 fractional factorial design with 8 factors in 16 trials. In such a design, the main effects are confused with three-factor interactions and the two-factor interactions are confused with each other. The chapter on design of experiments described the logic and construction of such experiments.

Noise Factor Design This aspect of the designed experiment was not considered in the chapter on design of experiments, so let us discuss it in some detail here. Each trial of the control factor design can be thought of as a prototype design of the system. From a robustness viewpoint, we wish to know how robust each prototype is to noise. Let us suppose that ambient temperature was one of the four noise factors. If this were the only noise factor, then we can evaluate each prototype at three levels of ambient temperature (low, medium, and high). In this fashion, the coating thickness for each of the 16 prototypes can be determined over the anticipated range of ambient temperature. From the collected data, the average and standard deviation of coating thickness can be determined. A prototype for which the average coating thickness is on target and the variability of coating thickness is the smallest possible is the best prototype. Among the infinitely many possible prototypes, we have constructed 16 in this experiment. It is therefore very unlikely that the best prototype is one of the 16 that we have constructed. To predict the control factor levels for the best prototype we could construct two equations as we did in Chapter 3, one for the average thickness $\bar{y}$ and another for the standard deviation of thickness s. The equations could be solved simultaneously to obtain the optimum levels of the control factors where the average coating thickness is on target and the variability of coating thickness is the smallest.

What if there are four noise factors instead of one. We want to evaluate the performance of each prototype over the entire range of noise. One way to accomplish this is to construct a designed experiment in just the four noise factors. This could be the eight-trial fractional factorial design in four factors. Each prototype could be evaluated under these eight noise conditions and the results analyzed as described above. The size of the experiment will become much larger, requiring 128 (16×8) evaluations. Is there a way to reduce the size of the experiment?

The size of the experiment can be dramatically reduced if we understand the general behavior of the noise factors. Consider the example of frozen pizza. Bake temperature and bake time are two important noise factors, because they cannot be controlled by the frozen pizza manufacturer and yet they have a large effect on the quality of baked pizza. Frozen pizza comes with a set of directions. It may say to bake the pizza at 425 degrees for 15 minutes. Depending upon the oven, when the dial is set to 425 degrees, the real temperature inside the oven varies from one consumer to the other and could be anywhere between 410 and 440 degrees. Also, it is a very rare individual that watches the pizza for 15 minutes! Bake time could vary between (say) 14 and 16 minutes. This defines the bake temperature and bake time domain of interest. We want the pizza to turn out to be good anywhere in this region. So, a noise factor experiment could be designed with three levels of temperature (410, 425, and 440 degrees) and three levels of time (14, 15, and 16 minutes) for a total of nine noise combinations. However, we understand baking to the extent that if the pizza is going to be limp that will occur at the low-temperature and low-time combination. If it is going to get burnt, that will happen at the high-temperature and high-time combination. If the pizza works well at these two combinations, it will work well at the remaining combinations. This means that the number of noise combinations can be reduced from nine to two, the low-time low-temperature combination and the high-time high-temperature combination. We do not need to know "everything you ever wanted to know about noise." We simply want to know the directional effects of noise. If this knowledge exists, the number of necessary noise combinations can be dramatically reduced. If such knowledge is not available, then an experiment in just the noise factors should be performed first to gain the knowledge necessary to be able to set up a small number of noise factor combinations that span the anticipated range of noise.

Table 4.1 assumes that the directional effects of noise are known. N_1, N_2, and N_3 represent three joint levels of the four noise factors that cover the entire noise domain. N_2 is the nominal combination of noise factors and N_1 and N_3 represent the range of noise of practical interest.

The following considerations apply in selecting the noise factors and their levels:

1. The design is made robust only against the selected noise factors and other noise factors that are related to those selected. The various types of noise factors described earlier should be well thought through to include the most important in the experiment.

2. If the directional effects of the noise factors are unknown, then an experiment in noise factors alone may have to be conducted first to gain the understanding necessary to construct a small number of joint levels for multiple noise factors.

3. If noise factors are unknown and replicate variability is very large, then that itself may serve as a measure of noise as was done with the stent coating experiment in Chapter 3. The number of replicates at each combination of control factors will necessarily be large to properly experience replicate variability.

4. The levels of noise factors, whether individual factors or joint factors, should be selected to cover the entire range of noise likely to occur in practice.

Analysis   As shown in Table 4.1, for each of the 16 trials, coating thickness was experimentally determined at the three noise conditions. For each trial, the average $\bar{y}$ and the standard deviation s calculated from the three observed thickness values are reported in Table 4.1. The standard approach to the analysis of designed experiments, described in Chapter 3, can be used to build equations for $\bar{y}$ and s. For the data in Table 4.1, these equations are

$$\bar{y} = 49.8 + 3.1A + 0.8B - 4.8C + \text{residual} \tag{4.1a}$$

$$s = 17.2 - 3.2A + 6.0B - 1.5C - 2.8AB + \text{residual} \qquad (4.1b)$$

Note that the same factors appear in both equations. The equations need to be *jointly* solved to find the values of A, B, and C such that the average coating thickness is on target $= 50$ and the standard deviation of coating thickness is minimized. This joint optimization could be accomplished purely mathematically, by plotting response surface contours, or by delegating the job to a software package with appropriate optimization routines.

This problem of joint optimization could have been decoupled if there were a factor, say factor C, that influences the mean but not the standard deviation. In such a case, we could find the values of control factors A and B that minimize the standard deviation from Equation (4.1b), substitute the values in Equation (4.1a), and then change factor C to get the mean on target.

Taguchi (1987) proposes that instead of analyzing the standard deviation, we should analyze the ratio $\bar{y}/s$, which is the inverse of CV. This is called the signal-to-noise ratio (S/N ratio), $\bar{y}$ is the signal and s is the noise. The S/N ratio is typically written in db units as $20\log(\bar{y}/s)$. As shown in Table 4.1, the S/N ratio can be calculated from the observed $\bar{y}$ and s values, and then analyzed using standard model building procedure to obtain its relationship with the control factors. The mean and S/N ratio equations for the data in Table 4.1 are

$$\bar{y} = 49.8 + 3.1A + 0.8B - 4.8C + \text{residual} \qquad (4.2a)$$

$$20\log\left(\frac{\bar{y}}{s}\right) = 10 + 1.8A - 2.9B + 0.9AB + \text{residual} \qquad (4.2b)$$

Note that factor C appears in the mean equation, but not in the S/N ratio equation. We can now find the values of A and B that maximize the S/N ratio. It is easy to see that, within the range of the experiment, the ratio is maximized when $A = 1$ and $B = -1$. For any specified average value of the response, maximizing the S/N ratio is the same as minimizing the standard deviation. These values of A and B are substituted in the mean equation and the value of C is determined such that the mean thickness is equal to the target of 50. This happens when $C = 0.44$. Because the joint optimization problem is now decoupled to make sequential optimization possible, we are able to find the optimum values of A, B, and C almost by inspection of Equations (4.2a) and (4.2b).

Was this decoupling of the joint optimization problem using the mean and S/N ratio equations, that is, finding an adjustment factor like C that influences the mean but not the S/N ratio, a fluke or is it likely to happen often? As an illustration, consider the volume knob of a stereo system. If we increase the volume knob, the mean volume increases and so does the standard deviation. Volume knob amplifies the music and also the noise. It leaves the ratio of mean to standard deviation unchanged. So, volume knob will be a factor in the mean equation, in the standard deviation equation, but not in the S/N ratio equation. This suggests that there are likely to be many factors that influence the mean and the standard deviation, but few that influence the S/N ratio. It is not easy to change the S/N ratio and that is why good stereo systems cost more. Decoupling is likely to be easier using the mean and S/N ratio equations.

Whether we should use mean and standard deviation equations, or we should use mean and S/N ratio equations need not be debated. Both approaches can be easily tried and the problem solved by using the approach that turns out to be simpler.

4.3   SIGNAL-TO-NOISE RATIOS

The signal-to-noise ratio to be used in analyzing a robust design experiment is a function of the type of response or characteristic of interest. For most common practical situations, the response may belong to one of the following types: smaller the better, nominal the best, larger the better, and fraction defective. We now consider the S/N ratios used for these common situations. The S/N ratios are best understood in the context of the quadratic loss function described below.

Quadratic Loss Function   Whenever a product characteristic deviates from the target, there is a loss sustained by the customer. The purpose of robust design is to minimize customer loss.

The economic loss function selected to quantify customer loss as a function of the deviation of the characteristic from target should have the following desirable properties. When the characteristic is on target T, the average economic loss to the customer can be assumed to be zero without loss of generality. If the characteristic is near the target, the economic loss should be small. As the deviation increases, the loss should increase *more than proportionately*. The quadratic loss function shown in Figure 4.4 is the simplest function with these desirable properties (Taguchi, 1987). When the product characteristic has a value y immediately after manufacture, the

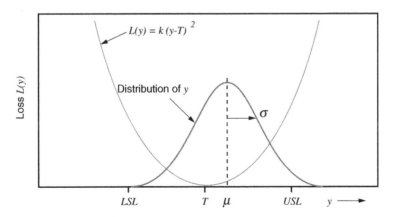

FIGURE 4.4   Economic loss function.

lifetime economic loss averaged over all customers is denoted by $L(y)$ and is given by the following quadratic loss function. The proportionality constant k is estimated by defining a single point on the loss function.

$$L(y) = k(y{-}T)^2 \qquad (4.3)$$

If the characteristic has a normal distribution with mean μ and standard deviation σ, as shown in Figure 4.4, then the loss caused by the mean not being on target and the variability not being zero can be obtained by integrating the quadratic loss function over the normal distribution. This average loss is

$$\text{Average loss} = k\left[(\mu{-}T)^2 + \sigma^2\right] \qquad (4.4)$$

This loss to the customer can be reduced by bringing the mean on target and by reducing variability. Note that when the mean is on target, the customer loss is directly proportional to the variance. This provides us with a powerful reason to reduce variability. It means that through your efforts if you reduce the variance to half, the pain and suffering of your customers is reduced to half!

The commonly used S/N ratios are described below in the context of the quadratic loss function.

Smaller the Better Characteristic   Impurities and pollution are examples of smaller the better characteristics. Let y_i be the observed values of the characteristic measured on a continuous scale. The characteristic can only take positive values. Zero is the target. For such characteristics, an adjustment factor that influences the mean without influencing the S/N ratio does not exist. Substituting $T=0$ in Equation (4.3), the customer loss is minimized when $\sum y_i^2 / n$ is the smallest possible. This is the same as maximizing the S/N ratio

$$\text{S/N ratio} = -10 \log\left(\frac{1}{n}\sum y_i^2\right) \qquad (4.5)$$

Nominal the Best Characteristic   Coat weight for a drug-coated stent is an example of a nominal the best characteristic. The characteristic is continuous and nonnegative. The target value is nonzero. For such characteristics, the variance becomes zero when the mean is zero. Also, usually an adjustment factor exists that can serve to adjust the mean without affecting the S/N ratio defined below. The rationale for this S/N ratio was explained in the previous section.

$$\text{S/N ratio} = 20 \log\frac{\bar{y}}{s} \qquad (4.6)$$

We first maximize the S/N ratio and then use the adjustment factor to bring the mean on target without affecting the S/N ratio. With mean on target, maximizing the S/N ratio implies minimizing the standard deviation. This leads to the minimization of customer loss.

Larger the Better Characteristic Pull strength is an example of larger the better characteristic. The characteristic is continuous and nonnegative. The target value is infinity. This problem can be restated as a smaller the better characteristic for $1/y$. Hence, the S/N ratio is

$$\text{S/N ratio} = -10 \log \left(\frac{1}{n} \sum \frac{1}{y_i^2} \right) \tag{4.7}$$

Fraction Defective This is the case of attribute data where fraction defective, denoted by p, can take values between 0 and 1. The best value for p is zero. To produce one good product, $1/(1-p)$ products have to be produced and the loss is equivalent to the cost of producing $\{(1/1-p)-1\} = p/(1-p)$ products. Hence, the S/N ratio is

$$\text{S/N ratio} = -10 \log \left(\frac{p}{1-p} \right) \tag{4.8}$$

4.4 ACHIEVING ADDITIVITY

Let us now turn to the second important objective of the robust design method, namely, to ensure that the design that is found to be the best in the laboratory continues to be the best in manufacturing and customer usage. This is accomplished by reducing control factor interactions, that is, by achieving additivity. Three types of interactions can occur between factors—control-by-noise (CN), noise-by-noise (NN), and control-by-control (CC). As explained before, the CN interactions are the key to robustness. The NN interactions can exacerbate or reduce the effects of noise factors. Detailed information on the main, interactive, and quadratic effects of noise is not necessary to design robust products. Directional information is sufficient to set up a joint noise factor that spans the noise domain. While CC interactions can prove useful when the control factor is also a noise factor, the following additional complications introduced by CC interactions should be considered:

1. If there are a large number of control factor interactions, the size and cost of the experiment will increase dramatically, the model will be very complicated, and the system will be difficult to understand.
2. We can think of the conditions of the experiment (laboratory, pilot, manufacturing, and customer usage) as control factors. As the product moves from the laboratory to manufacturing to customers, if a large number of control factor interactions are encountered at the laboratory stage, it is more likely that the control factors will also interact with the scale-up and usage factors. The knowledge gained in the laboratory experiments will not be transferable to downstream conditions.

Additivity means reducing or eliminating interactions between control factors. If such a result could be achieved, then the system will behave in a very simple fashion and scale-up will become easier. There are two ways to reduce CC interactions.

1. Unnecessary control factor interactions can be created by the wrong choice of responses. Every effort should be made to avoid this situation as explained below with examples.
2. There may be *a priori* knowledge that some control factors will interact. For example, force = mass × acceleration means that the interaction of mass and acceleration determines force. Also, if control factors have the same action mechanism, they will interact. As explained below, such known interactions can be eliminated in the designed experiment by a proper choice of control factors and their levels.

Selecting the Response Consider a pulverizing machine. The machine has three continuous control factors A, B, and C that sequentially reduce particle size. A 20–40-mesh particle size is desired, and the yield obtained in a one factor at a time experiment is shown in Table 4.2. The conclusion from this experiment is that factor B has the largest effect on yield followed by factor A. C has no effect on yield. Since A_2 is better than A_1 and B_2 is better than B_1, the best condition appears to be $A_2B_2C_1$ or $A_2B_2C_2$ with an expected yield of well over 90 percent. If the actual yield at the predicted optimum condition $A_2B_2C_2$ were to be less than 10 percent, it would be a surprise. If the prediction fails, it would suggest that the system behaves highly interactively, because the prediction was made without considering interactions (Taguchi, 1987).

For the same one factor at a time experiment, Table 4.3 shows the particle size data in terms of small, medium, and large particles. The 20–40-mesh data are the same as

TABLE 4.2 Yield as Response

	Trial	Percent Yield (20–40 Mesh)
A_1	(B_1C_1)	40
A_2	(B_1C_1)	80
B_2	(A_1C_1)	90
C_2	(A_1B_1)	40

TABLE 4.3 Particle Size as Response

		Particle Size		
	Trial	<20 Mesh	20–40 Mesh	>40 Mesh
A_1	(B_1C_1)	60	40	0
A_2	(B_1C_1)	0	80	20
B_2	(A_1C_1)	0	90	10
C_2	(A_1B_1)	0	40	60

those reported in Table 4.2. Looking at the effects of the three factors on the particle size distribution, it is clear that factor C is the most important factor because it pulverizes the product the most. The order of importance of effects is C first, then A, and then B. This is exactly the opposite of what was concluded from looking at yield alone in Table 4.2. It is also clear that the combination $A_2B_2C_2$ will result in close to 100 percent dust or almost 0 percent yield. Had we looked at Table 4.3 before, we would never have suggested $A_2B_2C_2$ as the optimum combination. We would have realized that in going from B_1 to B_2 we pulverized the product too much and perhaps a level such as $A_1B_{1.8}C_1$ should be investigated. The observed yield will exceed 90 percent as expected. This success of the confirmation experiment will mean that the system is noninteractive because the result of the confirmation experiment was predicted without considering interactions and the prediction came true.

The message is that when we look at the percent yield as a response, the system behaves interactively. When we look at particle size distribution as the response, the system behaves noninteractively. Thus, by selecting the right response, control factor interactions can be reduced. In general, percent yield defined as the percent of the product within a *two-sided* specification, while it is an important measure from a business point of view, is the wrong response to guide R&D. The responses need to be carefully selected to ensure that they do not engender unnecessary control factor interactions.

Selecting Control Factors and Levels   Consider a multistage drier used to drive off moisture. Suppose that the temperature and residence time in each stage are used as control factors. Then, for a three-stage drier, there will be six control factors. It is easy to see that all these control factors will interact with each other. For example, the effect of the second-stage temperature will depend upon how high or low the first-stage temperature was. The six control factors interact because they serve the same function of supplying energy. In selecting control factors, it is important that each control factor influence a distinct aspect of the basic mechanism of the system. Otherwise, the control factors will interact. In such a case, only one control factor from the group may be selected, or the levels of one control factor may be made a function of the level of the interacting control factor as follows.

Time and temperature often interact because it is the energy input that matters. There is an optimum energy input that can be obtained with high-time and low-temperature combination or with low-time and high-temperature combination. Thus, the effect of time can be positive at low temperature and negative at high temperature, causing an antisynergistic interaction between the two factors. One way to reduce this interaction is by selecting the right response—as was explained earlier. For example, in a baking experiment, instead of measuring whether the bread is acceptable or not, the color of the bread could be measured. A second way to reduce the interaction is by making the levels of time a function of temperature. If 10 minutes is a reasonable midpoint at 400 degrees, 20 minutes may be reasonable at 300 degrees. The two levels of time could be taken to be ±20 percent from the midpoint. This judgment has to be based upon technical understanding and experience. To the extent the judgments are right, the time temperature interaction will reduce dramatically.

4.5   ALTERNATE ANALYSIS PROCEDURE

Any proposed analysis procedure needs to be robust, that is, it should not lead the experimenter astray even if mistakes are made in the design of the experiment. This is not to say that experimenters should not think deeply before designing the experiment, but to recognize that there are limits to prior knowledge. As an illustration of this idea, it was demonstrated in Chapter 2 that confidence interval for the difference of means is a much more robust analysis procedure compared to a t-test because confidence interval does not lead to wrong conclusions even if the sample size is wrong. The nominal-the-best S/N ratio analysis procedure is now further examined to show that it can lead to a wrong optimum, can artificially generate interactions and quadratic effects, and with highly fractionated control factor designs, can identify wrong control factors as being responsible for achieving robustness (for further examples and references, see Steinberg and Bursztyn (1994)). These limitations extend to several other S/N ratios, and apply equally well to other proposed summary statistics such as standard deviation, log(standard deviation), CV, and log(CV), with and without data transformation. The limitations arise because noise factors are treated as random factors rather than as fixed factors in these analyses of summary statistics. In some cases, these limitations can be overcome by building an equation for data as a function of the control and noise factors, that is, by explicitly modeling control-by-noise interactions. On the other hand, there are arguments in favor of the S/N ratio analysis as well so that both types of data analyses may be conducted with conclusions to be confirmed by a confirmation test.

S/N Ratio Identifies the Wrong Optimum   The data for a very simple experiment are shown in Table 4.4 to illustrate how the nominal-the-best S/N ratio $= 20 \log(\bar{y}/s)$ can lead to a wrong optimum. In Table 4.4, A is a continuous control factor, N is a noise factor, and y is the response. All effects are expected to be linear.

In Table 4.4, the average $\bar{y}$, the standard deviation s, and the S/N ratio are the same at the (-1) and $(+1)$ levels of control factor A. The conclusion is that control factor A has no effect on all these responses. In particular, the conclusion is that A cannot be used to minimize the standard deviation or to maximize the S/N ratio, that is, A cannot be used to improve robustness.

Figure 4.5 shows a plot of the data. Clearly, control factor A and noise factor N interact, and A can be used to achieve robustness. The equation for data y as a function of A and N is

$$y = 20 - 10AN$$

TABLE 4.4   A Simple Robustness Experiment

A	$N = -1$	$N = +1$	$\bar{y}$	s	S/N ratio $= 20 \log(\bar{y}/s)$
-1	$y = 10$	$y = 30$	20	14	3.1
$+1$	$y = 30$	$y = 10$	20	14	3.1

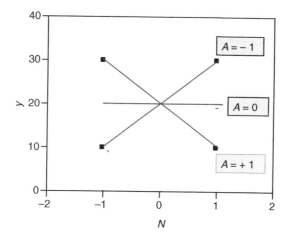

FIGURE 4.5   Plot of AN interaction (negative interaction).

Since A is a continuous control factor, $A = 0$ is expected to make the effect of N disappear. Explicit modeling of control-by-noise interactions shows that control factor A can be used to achieve robustness and that the optimum level of A is $A = 0$. The S/N ratio analysis failed to identify this optimum.

This failure of the S/N ratio to identify the optimum has nothing to do with whether the AN interaction between the control factor A and the noise factor N is negative or positive. Consider the following experiment in Table 4.5 and analysis.

Data analysis leads to the following two equations:

$$\bar{y} = 16.5 + 3.5A$$

$$\text{S/N ratio} = 11.25 - 8.15A$$

The S/N ratio equation suggests that A should be made as small as practically possible. One would then use another factor B (not discussed here), which influences the average but not the S/N ratio, to get the average on target.

Figure 4.6 shows a plot of the data. The equation for data y is:

$$y = 16.5 + 3.5A + 5.5N + 4.5AN = 16.5 + 3.5A + (5.5 + 4.5A)N$$

To make the effect of N disappear, we must have $5.5 + 4.5A = 0$, leading to the optimum value of $A = -1.22$. Again the S/N ratio equation fails to find the optimum.

TABLE 4.5   An Experiment with Positive AN Interaction

A	$N = -1$	$N = +1$	$\bar{y}$	s	S/N ratio $= 20\log(\bar{y}/s)$
-1	$y = 12$	$y = 14$	13	1.4	19.4
$+1$	$y = 10$	$y = 30$	20	14.0	3.1

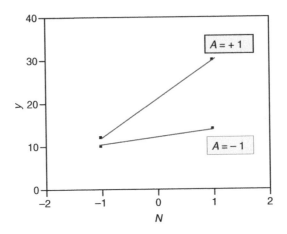

FIGURE 4.6   Plot of *AN* interaction (positive interaction).

S/N Ratio Induces Control Factor Interactions   Not surprisingly, the difficulties with using a summary statistic such as an S/N ratio can multiply as the experiment becomes more complicated. In particular, the use of S/N ratio can induce control factor interactions and with resolution 3 control factor experiments, these induced interaction effects can misleadingly appear as the effects of certain control factors. This may be seen as follows from a simple illustrative experiment. In the following, A, B, and C are continuous control factors and N is a noise factor. The results are as summarized in Table 4.6.

Suppose we want the mean response y to be on target $=10$ with the smallest variance. The results seem obvious by inspection. The usual S/N ratio analysis will lead to:

$$\bar{y} = 10 + 2A$$
$$\text{S/N ratio} = \text{maximum when } C = -1$$

The conclusion is that the control factor A influences the average, control factor C influences the S/N ratio, and the objectives will be met if $C = -1$ and $A = 0$.

The entire experiment can be analyzed to fit an equation to the data to include the effects of control and noise factors and control-by-noise interactions.

TABLE 4.6   S/N Ratio Identifies the Wrong Control Factor

A	B	C	N=−1	N=+1	$\bar{y}$	s	S/N ratio $= 20 \log(\bar{y}/s)$
−1	−1	1	$y=10$	$y=6$	8	2.8	Small
−1	1	−1	$y=8$	$y=8$	8	0	Large
1	−1	−1	$y=12$	$y=12$	12	0	Large
1	1	1	$y=10$	$y=14$	12	2.8	Small

$$y = 10 + 2A + AN + BN = 10 + 2A + (A+B)N$$

It follows that

$$\bar{y} = 10 + 2A$$
$$\sigma_y = |A+B|\sigma_N$$

The conclusion is that the control factor A influences the average, but the lack of CN interaction in the equation for y means that control factor C cannot be used to achieve robustness. Robustness will be achieved whenever $(A+B) = 0$, for example, when $A = 1$ and $B = -1$, $A = 0$ and $B = 0$, and so on.

In this case, the S/N ratio analysis identified a wrong control factor C as the source of robustness. The reason may be understood as follows. The effect of the noise factor N on y is large when the absolute value of $(A+B)$ is large and vice versa. For the four combinations of A and B in Table 4.6, the absolute values of $(A+B)$ are 2, 0, 0, and 2, respectively, and these perfectly correlate with the values of AB interaction 1, -1, -1, and 1. Thus, the standard deviation of y and, therefore, the S/N ratio become a function of AB interaction. Since the control factor experiment in Table 4.6 is a resolution 3 experiment with the confounding $C = AB$, the S/N ratio artificially turns out to be a function of factor C.

In Section 4.4, it was pointed out that responses should be carefully selected so that they do not create artificial control factor interactions. The S/N ratio itself can be a response that can create artificial control factor interactions.

Limitations of Other Summary Statistics   The limitations cited above for the nominal-the-best S/N ratio also apply to other summary statistics such as the standard deviation, log(standard deviation), CV, and log(CV) as can be easily seen from Table 4.4. It is also easy to see from this table that the limitations will carry over to S/N ratios for smaller-the-better, larger-the-better, and signed target characteristics.

Why These Limitations?   These limitations occur because noise factors, which are fixed factors in the experiment, are treated as random factors in the S/N ratio analysis. The analysis assumes that summary statistics such as standard deviation or S/N ratio are sufficient statistics to describe all the robustness information contained in the experiment in the form of control-by-noise factor interactions. This thinking is incorrect. If the observations corresponding to any specific trial of the control factor experiment are shuffled, the summary statistics (S/N ratio, standard deviation, etc.) will not change, but the control-by-noise factor interactions can change dramatically. This lack of sufficiency is the reason why the S/N ratio analysis led to the wrong conclusions in the examples cited above.

Alternate Analysis Procedure   Instead of using the average and the S/N ratio, another approach is to first build an equation for response y as a function of control and noise factors and their interactions. For example, the equation for three control factors A, B, and C and two noise factors N_1 and N_2 can be:

$$y = a_0 + a_1 A + a_2 B + a_3 C + a_4 AB + a_5 N_1 + a_6 N_2 + a_7 AN_1 + a_8 AN_2 + a_9 BN_1$$

The following optimization procedure can be used.

1. Group the coefficients for each noise factor together and set them to zero.

$$a_5 + a_7 A + a_9 B = 0$$

$$a_6 + a_8 A = 0$$

The two equations will be solved simultaneously to get the solutions A^* and B^*.

2. The equation for average is

$$\bar{y} = a_0 + a_1 A + a_2 B + a_3 C + a_4 AB$$

Substitute A^* and B^* in this equation for average and use a factor similar to C that influences the average but does not interact with the noise factors to bring the average on target. The probability of finding a factor similar to C could be high if many control factors influence the response but few control factors interact with the noise factors.

This analysis procedure is likely to work well when strong noise factors are identified and the noise factor array is a fractional factorial design or a small number of joint noise factor levels.

It is possible that depending upon the complexity of the equation for response y, the optimization procedure could get more involved. For example, a factor similar to C that influences the average but does not interact with the noise factors may not be found. Perhaps it could be that levels of control factors that lead to robustness against one noise factor are different from the levels of control factors that lead to robustness against another noise factor. The necessary trade-off could be made in various ways including the creation of two equations, one for the average and the other for the variance (using variance transmission analysis) based upon the fitted equation for the response y.

Reasons for Using S/N Ratio Analysis   There are reasons to prefer the S/N ratio analysis as well. These include the following:

1. The S/N ratio analysis procedure is simple to execute.
2. There are situations when, instead of an explicit consideration of noise factor levels, the effect of noise is incorporated by many random replicates as was done in the drug-coated stent example in Section 3.6. In this case, the control-by-noise interactions cannot be computed and the summary statistics approach of using the S/N ratio or log s is the way to proceed.
3. In addition to the noise factors explicitly considered in the experiment, there usually are other known or unknown noise factors that also influence the response. The data in any row of the matrix experiment vary not only because of

the changes in identified noise factors, but also due to all other factors captured by replication. Thus, the effect of noise factors not included in the experiment will be captured by the summary statistics such as the standard deviation and the S/N ratio but not by the model for response y (except as residual). When factors other than the identified noise factors dominate variability in the rows of the matrix experiment then the situation becomes similar to that of random replication and the S/N ratio or log s analysis will be preferred. Alternatively, the model for y may be used to assess robustness against the identified noise factors and the S/N ratio analysis can then be used on the residuals of this model to assess robustness against replication. However, if the number of identified noise factor combinations is small, the effects of excluded noise factors are less likely to be properly captured and could result in wrong conclusions. The idea of explicitly incorporating major noise factors and varying their levels widely in the matrix experiment is intended to avoid reliance on a small number of random replicates.

4. It is possible that the underlying true relationship between the response and the control and noise factors is complex and not well captured by the fitted model for y. In this case, the model inadequacy will appear as residuals in the fitted equation for y, and may be better captured by the average and S/N ratio approach.

Confirmation Test   With highly fractionated experiments and alternate analysis procedures, it is prudent to insist upon a conformation test(s). The experiment should be analyzed both ways: using the S/N ratio analysis and the equation for the response y as a function of control and noise factors and their interactions. If both analyses suggest similar conclusions and optimum, the predicted optimum should be confirmed by a confirmation test.

Otherwise, the confirmation tests need to be judiciously selected. If the two analyses predict two separate optima, one possibility is to select the confirmation tests as the two predicted optima. To illustrate, let us reconsider the analysis of data reported in Table 4.6. The S/N ratio conclusion was to use $C = -1$ regardless of the levels of A and B. The conclusion from the analysis of the response y was to use levels of A and B such that $(A + B) = 0$ regardless of the level of C.

Had we only done the S/N ratio analysis, it would have been tempting to suggest the confirmation experiment as $(A = -1, B = +1, C = -1)$ or $(A = +1, B = -1, C = -1)$. Both these experiments are already done in the initial designed experiment and could have wrongly confirmed the S/N ratio conclusion.

Now that the analysis has been done in two ways, and has led to completely different conclusions, the confirmation tests could be judiciously structured to disprove the conclusions. For example, the two confirmation tests could be as follows:

$(A = +1, B = +1, C = -1)$ to disprove the S/N ratio conclusion

$(A = 0, B = 0, C = +1)$ to disprove the conclusions from the equation for y

4.6   IMPLICATIONS FOR R&D

The robust design methodology described above requires some significant changes to the way R&D is conducted in many companies.

Table 4.7 shows the various stages of R&D and manufacturing and the type of noise that can be counteracted at each stage (Taguchi, 1987). At the product design stage, the effects of all types of noise can be reduced by the following three activities:

1. *System Design:* Knowledge of science and technology is necessary to think about widely different ways of designing a product to meet the desired function. Design of experiments and robust design approaches do not play a major role at this stage. To transfer drug through skin, a patch is one possible drug delivery system. Another way to deliver drug through skin may be to use a constant-current device where the drug transfer rate is proportional to the current, and current can be kept constant using an appropriate circuit design. For such a system, the effect of skin-to-skin differences will disappear (although at a higher cost). System design requires specialized technical knowledge and is the domain of the scientists and engineers.

2. *Parameter Design:* Parameter design has the potential to counteract all types of noise. Once a specific system is selected, we need to determine the target values for the control factors (for a patch—types of ingredients, percentage of each ingredient, thickness of the patch, etc.) in such a way that the outputs are at their target values and the effect of noise factors on outputs is minimized. This is what the robust design approach is meant to accomplish.

3. *Tolerance Design:* This is the development of ranges for control factors such that within the selected ranges the outputs meet their specifications. Factors that have the largest effects on outputs will have narrow ranges. Tighter tolerances usually imply higher cost of the product. That is why parameter design is done first to permit the widest possible tolerance. This subject of setting tolerances or specifications is discussed in Chapter 5.

TABLE 4.7   Stages of R&D and Manufacturing and Ability to Reduce Effects of Noise Factors

Stage	Activity	External	Degradation	Manufacturing Variability
Product design	System design	Yes	Yes	Yes
	Parameter design	Yes	Yes	Yes
	Tolerance design	Yes	Yes	Yes
Process design	System design	No	No	Yes
	Parameter design	No	No	Yes
	Tolerance design	No	No	Yes
Manufacturing	Process control	No	No	Yes
	Screening	No	No	Yes
Customer usage	Warranty and repair	No	No	No

The activities of system design, parameter design, and tolerance design also apply at the process design stage. At the process design stage, we cannot reduce the effects of either the external noise factors or factors that cause product deterioration. This can be done only at the product design stage. However, manufacturing variability can be reduced at the process design stage by reducing the effect of noise factors that affect manufacturing variability (system and parameter design) and by setting specifications for process factors (tolerance design).

During manufacturing, engineering feedforward and feedback controls and statistical process control are used to detect and correct drifts of the manufacturing process. Screening by acceptance sampling plans or 100 percent inspection becomes the last step to assure quality.

If poor quality product is shipped to the customer, the only thing left to do is to provide service and compensate the customer in an effort to prevent further loss to the reputation and market share of the manufacturer.

The basic principles of robust design and the feasibility of quality countermeasures summarized in Table 4.7 have major implications toward how products and processes should be designed and scaled up. These basic principles may be summarized as follows:

1. The robustness of a product can be increased only if there are interactions between control and noise factors and these interactions are exploited during the product and process design stages.
2. Interactions between control factors are detrimental to product scale-up and efforts should be made to reduce such interactions.

These principles require major changes to the way products are designed and scaled up. In many companies, R&D is done by treating noise factors as nuisance factors and every effort is made to ensure that experiments are not affected by noise. Thus, R&D has the best equipment to minimize any variation due to processing. Perhaps only a single batch of raw material is used to minimize the impact of raw material variation. Product evaluations are done by trained personnel to minimize any impact of customer factors. Every effort is made to *control noise* as research and development proceeds. After the product design is complete, it is *evaluated* for robustness with the hope that it turns out to be robust. If not, costly redesign or excessive manufacturing controls or restrictions on customer usage conditions become necessary causing delays and increase in cost.

What is needed instead is an explicit consideration of noise factors early in the R&D phase, and experiments with noise factors to reduce the sensitivity of product design to noise. Thinking about factors that will change during scale-up and those that will become noise factors in manufacturing (even though they could be well controlled during R&D), including manufacturing engineers in the development team, making the development team responsible to transition and manage early production are all ways to increase timely consideration of noise factors. Even though noise factors such as ambient temperature or skin cannot be controlled during manufacturing and customer usage, laboratory experiments can be conducted by holding the temperature or cadaver skin at the desired levels. Such experiments need to be conducted early in the product design phase to build robust products. Also, greater technical thinking should be promoted to properly select the responses, S/N ratios, control factors, and their levels to reduce control factor interactions.

4.7   QUESTIONS TO ASK

Here are some questions to ask in addition to the questions listed in the chapter on design of experiments. The questions listed below are specifically related to the use of noise factors to build robustness and considerations to achieve additivity.

1. Are noise factors being considered early in the product and process design stage or is R&D being conducted keeping noise factors as constant as possible?
2. Is an appropriate cross-functional team involved in selecting the likely noise factors?
3. Are all the potentially important noise factors being identified? These include external noise factors, product degradation factors, and factors producing manufacturing variability.
4. Is there enough knowledge regarding the directional effects of noise factors to be able to construct a small number of joint noise factor levels that will span the noise space of interest? If not, is an experiment in noise factors alone being planned to generate such knowledge?
5. Are responses being selected to ensure that they do not unnecessarily generate control factor interaction?
6. Are the control factors and their levels being selected to reduce control factor interactions?
7. Is the robustness experiment correctly structured to be able to estimate control factor by noise factor interactions?
8. Are proper signal-to-noise ratios being used to analyze the experiment?
9. If noise factors are unknown and replicate variability is large, are efforts being made to determine the control factors that affect replicate variability?
10. Is all available scientific and engineering knowledge being fully utilized?

Further Reading   Genichi Taguchi developed the robust design method and this chapter is based upon his work. Taguchi (1987) covers the robust design method in detail and also provides a large number of practical examples from different industries. Phadke (1989) is easier to understand and has several case studies and a particularly good explanation of the signal-to-noise ratios and dynamic characteristics.

CHAPTER 5

SETTING SPECIFICATIONS: ARBITRARY OR IS THERE A METHOD TO IT?

Specifications for product, process, and raw material characteristics are often poorly set in industry. This chapter explains the R&D and manufacturing implications of specifications, and how to use the basic principles of empirical, functional, and life cycle cost approaches to set meaningful specifications.

Making Lemonade   Consider the question of developing and producing the best-tasting lemonade as measured by a consumer taste test. The degree of liking is measured in the food industry by a hedonic score on a seven-point scale with the score of seven being the highest possible. Suppose we decide that we need to ensure that our lemonade will always get a score of at least six. Then, hedonic score ≥ 6 will be the one-sided specification for hedonic score. Our lemonade consists of water, lemon juice, sugar, salt, and other ingredients. How should we set the targets and specifications for these ingredients? Several different approaches may be taken.

1. *Empirical Approach:* Suppose we have to set the specifications for percent lemon juice. Let us say that we have made five batches of lemonade in the past based upon our family recipe. All the batches received a hedonic score >6. Because we were a bit sloppy in making the batches of lemonade, the percent lemon juice varied somewhat from batch to batch. Luckily, we kept track of percent lemon juice in each batch. So we have six numbers for percent lemon juice. What we could do is to make the average $\bar{x}$ of these six numbers as our

Industrial Statistics: Practical Methods and Guidance for Improved Performance   By Anand M. Joglekar
Copyright © 2010 John Wiley & Sons, Inc.

target for percent lemon juice. Also, if future batches have percent lemon juice in a range suggested by the numbers we have observed, then the future batches of lemonade will also get a hedonic score ≥ 6. So, one possibility is to specify the observed range as the specification. We could also calculate the standard deviation s from the six observations and use $\bar{x} \pm 3s$ as the specification with the hope that 99.7 percent of the future batches will meet this specification. A third possibility is to use the tolerance interval $\bar{x} \pm ks$ as the specification with an appropriately selected value of k. These are examples of an empirical approach to developing specifications. Many a time, specifications in industry are developed using the empirical approach without having an explicit understanding of the functional relationship between inputs and outputs.

2. *Functional Approach:* This approach is based upon an understanding of the functional relationship between inputs and outputs, in our case, the relationship between the ingredients and the hedonic score. If the percent lemon juice is too low, people will not like the lemonade much (not lemony enough). As percent lemon juice is increased, the hedonic score will increase reaching a maximum. If the percent lemon juice becomes too high, again people will not like the lemonade (too lemony) and the hedonic score will drop. This relationship between percent lemon juice and hedonic score will be parabolic with a maximum. The relationship can be experimentally determined by purposefully making batches of lemonade with different percent lemon juice and getting the hedonic score for each batch. The percent lemon juice specification (target and range) can then be set such that the *hedonic score* is ≥ 6 within the specification. To set the specifications for all the ingredients, we must first find the equation that jointly relates these ingredients to the hedonic score. This can be done by conducting designed experiments. The functional approach uses this relationship to develop specifications. Perhaps the equation shows that if all the ingredients are exactly on target, the hedonic score is 6.5. But we know that from one batch to the next we cannot hold all ingredients exactly on target. If the specification for hedonic score is hedonic score ≥ 6, then each ingredient can be allowed to be within some acceptable range around the target such that the hedonic score will still meet its specification. This is how specifications are set based upon an understanding of the functional relationship.

3. *Cost-Based Approach:* If we can hold each ingredient in a very narrow range around the target, narrower than what was set using the functional approach, the hedonic score will be (say) >6.3 within this narrower range. However, it will cost more to ensure that all ingredients are always tightly controlled in this very narrow range. Is this extra expense justified? The answer depends upon what benefits will be received because of the 0.3 increase in the hedonic score. The greater the hedonic score, the greater the degree of liking, and perhaps the greater the sales volume. If the profits we make because of the added hedonic score are greater than the cost we incur in tightly controlling all ingredients, then we will be willing to accept the narrower specifications. This is how specifications are set based upon an understanding of the cost structure.

5.1   UNDERSTANDING SPECIFICATIONS

Specifications may be two-sided or one-sided. A two-sided specification has a target and a permissible range around the target. The drug content of a tablet may have a target value of 2 mg with a two-sided specification 2 ± 0.3 mg. At times, two-sided specifications are asymmetrical with different ranges on each side of the target. A one-sided specification does not have a target, or the target may be zero (smaller the better) or infinity (larger the better). The residual solvent of a tablet may have a one-sided specification with only an upper specification limit. The target is zero. Pull strength may have a one-sided specification with only a lower specification limit. The target is infinity. Let us first consider how such specifications are interpreted in practice.

What Do Specifications Mean?   One usual interpretation is that if the characteristic (e.g., drug content) is within the specification, the product is good, and if the characteristic is outside the specification, the product is bad. This interpretation is flawed. While it does result in efforts to produce products within specifications, the incentive to produce products exactly on target does not exist because the interpretation does not explicitly recognize the benefits of doing so.

A second interpretation of specifications is to first recognize that an on-target product (assuming that the target is correctly set) is the best. As drug content begins to deviate from target, even inside the specification, the effectiveness of the tablet begins to erode. Large deviations from target ultimately produce unacceptable deterioration in effectiveness (under-dose) or unacceptable side effects (overdose). It is not true that inside the specification the tablets are equally good. If you went to a store to purchase a TV set and found one marked "Right on target!" and another marked "Just at the edge of specification!," it is clear which one you will pick even though both are inside specification. Outside the specification, the tablets do not instantly turn into bad tablets. In fact, it will be virtually impossible to even distinguish between a tablet that is just inside the specification and another that is just outside. Specification is a line in the sand intended to prevent bad consequences. In some industries, specifications become contractual and legal obligations, but functionally, they are lines in the sand nonetheless.

The implications of these two differing interpretations toward the setting of specifications will be explored later.

What to Put Specifications On?   Everything that moves does not need to have a specification. Specifications should be developed only for those characteristics that have an impact on the final product performance.

Figure 5.1 shows a cascading set of characteristics. Consider a drug in the form of a suspension intended to be self-injected by the patient using a syringe. The patient expectation is to be able to accomplish the task easily. This expectation could be quantified as a performance characteristic called injection force, with a specification based upon patient feedback. This injection force is a function of product characteristics such as viscosity and drug particle size. Injection force is also a function of device characteristics such as needle diameter and syringe diameter. Drug particle

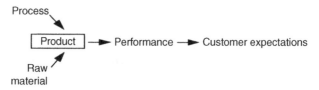

FIGURE 5.1   Cascading set of characteristics.

size is a function of process characteristics such as mix speed and mix time, and it is also a function of raw material characteristics. Specifications should be developed for these key characteristics because they control product performance.

Thus the job of developing specifications begins with setting performance (or product) specifications based upon customer expectations. Then, the key product, process, and raw material characteristics that control performance need to be identified. This is an important task for R&D and design of experiments is an important tool to accomplish this task. Specifications should be developed for these key characteristics such that if the key characteristics meet their specifications, the performance characteristic will meet its specification.

R&D and Manufacturing Implications   Once the specifications are set, whether externally mandated or internally developed, they imply that R&D and manufacturing must meet certain variability targets to manufacture products with low fraction defective.

Consider the usual situation of lot-by-lot production with each lot consisting of 1000 or 100,000 products as the case may be. Consider any one product characteristic such as drug content of a tablet. Drug content varies due to two reasons: not all tablets within a lot have identical drug content and not all lots have the same mean drug content. These are referred to as the within-lot and between-lot variance components (see Chapter 8). The total variance of drug content is the sum of these two component variances as shown in Equation (5.1) where σ_t is the total standard deviation, σ_w is the tablet-to-tablet standard deviation of drug content within a lot, and σ_b is the standard deviation of mean drug content between lots.

$$\sigma_t^2 = \sigma_w^2 + \sigma_b^2 \tag{5.1}$$

To a large extent, σ_w depends upon product and process design and this is the responsibility of R&D. To a large extent, σ_b has to do with control of the manufacturing process and this is the responsibility of manufacturing. What should σ_t be so that the products consistently mean specifications and how should it be divided between R&D and manufacturing? If R&D is allowed a large σ_w, manufacturing will be left with a small σ_b requiring tight process controls. If manufacturing is allowed a large σ_b, the job of product and process design in R&D will become more difficult to achieve a small σ_w.

If the P_p index (see Chapter 7) defined by Equation (5.2) is set to be 1.33, most of the products will be inside specifications assuming that the long-term process mean is near target. Let W represent the half-width of a two-sided specification or the distance between process mean and specification limit for a one-sided specification. Then,

$$P_p = \frac{W}{3\sigma_t} \tag{5.2}$$

If $P_p = 1.33$, then $\sigma_t = W/4$ and from Equation (5.1)

$$\frac{W^2}{16} = \sigma_w^2 + \sigma_b^2 \tag{5.3}$$

Equation (5.3) is the equation of a circle with radius $= W/4$ and is shown in Figure 5.2 for various values of W.

Let us consider the implications of Figure 5.2 in the context of an example. Suppose $W = 8$, then, the radius of the circle will be 2. This means that for $P_p \geq 1.33$, σ_w and σ_b

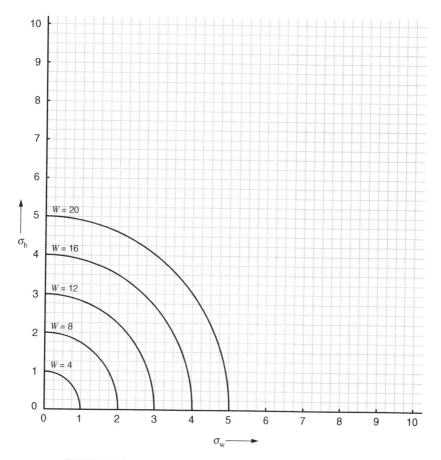

FIGURE 5.2 Impact of specification on σ_w and σ_b ($P_p \geq 1.33$).

will have to be inside the circle marked $W = 8$. For $\sigma_w = 2$, σ_b will have to be zero, which is not practical. Similarly, for σ_b equal to 2, σ_w will be zero, which is also not practical. As σ_w becomes less than 2, the permissible value of σ_b initially increases rapidly. Thus, $\sigma_w = \sigma_b = 1.4$ is possible. Further reductions in σ_w eventually lead to a diminishing increase in σ_b.

Figure 5.2 provides us a way to understand the implication of specifications toward setting variability targets (σ_w and σ_b) for R&D and manufacturing (see Joglekar (2003) for similar considerations for more complex specifications). The variability targets for R&D and manufacturing can be set based upon the constraints imposed by Figure 5.2 and the difficulty of accomplishing them in practice.

Let us now turn to the question of setting specifications assuming that the key characteristics have been identified. In general, targets and permissible ranges should be jointly set to permit widest possible ranges for key characteristics while achieving the desired performance index P_{pk} (see Chapter 7) for the output characteristic. This results in robust products and processes. Often in industry, the targets are already set, and then, the question regarding permissible ranges for key input factors is asked. This can result in suboptimization, as explained in Chapters 4 and 9.

There are three approaches to set specifications: empirical approach, functional approach, and life cycle cost approach. We begin with the empirical approach, which is widely used in the industry.

5.2   EMPIRICAL APPROACH

The moisture content of a product is believed to be an important functional characteristic. However, the functional relationship between moisture content and product performance is unknown. Ten acceptable batches were produced with the following observed percent moisture: 2.3, 2.4, 2.2, 2.3, 2.3, 2.1, 2.2, 2.3, 2.1, and 2.3. What should be the two-sided specification for moisture?

One approach is to say that the target should be the observed average 2.25 and the specification should be the range of observed data, namely, from 2.1 to 2.4, because the final product was acceptable in this range of moisture. However, a large number of future batches are likely to fail this observed range because the range is based upon a small sample size. Assuming a normal distribution for moisture and using tolerance intervals, it can be shown that we are 95 percent sure that only about 60 percent of the future batches will have a moisture content inside this specification. If the moisture specification is set to be from 2.1 to 2.4, approximately 40 percent of future production is likely to be rejected.

A second approach is to assume that the process that resulted in these observed moistures is an acceptable process, and then, base the moisture specification on our understanding of the moisture distribution. What is often done in industry is to first compute the sample average $\bar{x} = 2.25$ and the sample standard deviation $s = 0.097$, then construct an interval $\bar{x} \pm 3s (= 1.96\text{–}2.54)$ and use it as a specification in the belief that this interval will capture 99.7 percent of the future batches. This calculation turns out to be wrong. While $\mu \pm 3\sigma$ does capture 99.7 percent of the batches, $\bar{x} \pm 3s$ does not, since $\bar{x}$ and s are not μ and σ.

A third approach is to compute an interval that correctly captures the desired percentage of future batches (say) 99.7 percent. This is given by the tolerance interval $\bar{x} \pm ks$, where $\bar{x}$ is the target and $\pm ks$ represents the acceptable range. The value of k depends upon the sample size n, percent confidence, and percentage of the population to be included inside the interval. The values of k are tabulated in the Appendix or the interval can be computed using the appropriate software. For $n = 10$, 95 percent confidence, and 99.7 percent of the population, $k \approx 5.15$ (not 3) and the tolerance interval is 1.75–2.75. If this is defined to be the specification for moisture, we are 95 percent sure that 99.7 percent of the future production will be inside the specification if the process that generated the original moisture data remains unchanged.

Even with these tolerance interval calculations, the empirical approach suffers from the major drawback that it is not based upon a functional understanding of how moisture content influences product performance. It is possible that the empirically derived specification (1.75–2.75) may be unnecessarily too narrow or unacceptably too wide to achieve acceptable product performance. The empirical approach should be used only as a last resort.

5.3 FUNCTIONAL APPROACH

Functional approach to setting specifications is based upon an understanding of the functional relationship between the input characteristics and the output characteristic. Two questions arise: How to find the functional relationship and how to set specifications based upon the relationship? These questions are now answered. The impact of measurement variability on setting specifications can be, but has not been, considered in the discussion that follows.

Finding Functional Relationships A functional relationship may be an explicit equation, a set of partial differential equations, or a computer program. Let $y = f(x_i)$ represent the functional relationship between output y and inputs x_i. y may be a performance characteristic and x_i may be product characteristics, or y may be a product characteristic and x_i may be the process and raw material characteristics.

There are three ways to find functional relationships: mechanistic modeling, empirical mechanistic modeling, and empirical modeling. Chapters 3, 4, and 10 provide many examples of empirical models based upon designed experiments and regression. Chapter 10 also includes examples of how mechanistic models may be obtained. A brief summary is provided here.

Mechanistic models are based upon an understanding of the mechanism, namely, based upon theory. Simple examples are as follows:

- Drug content = fill weight × drug concentration
- Volume of a sphere = $(4/3)\pi r^3$, where r is the radius.

- For a Newtonian fluid, injection force as a function of syringe diameter (r), needle diameter (R), viscosity (v), needle length (L), and flow rate (Q) is given by

$$\text{Injection force} = \frac{kr^2vLQ}{R^4}$$

Other examples include equations used for electronic circuit design and fluid flow. A mechanistic model for release rate CV of a controlled release tablet is derived in Chapter 10. Mechanistic models are preferred because they provide greater understanding and are more predictive over wider ranges of input factors.

Empirical mechanistic models are based upon an incomplete understanding of the underlying mechanism. Perhaps there is enough of an understanding to propose a general form of the model, but the model coefficients are unknown and need to be estimated from data. For example, for non-Newtonian flow, the equation for injection force may look similar to the one given above for Newtonian flow, but with different unknown exponents to be estimated from data. Chapter 10 shows how this can be done efficiently.

Empirical models are equations derived purely based upon data. Regression analysis and design of experiments are useful statistical tools to derive these equations expressing an output as a function of input factors. These equations are valid over the ranges of input factors used to generate the data. Their prediction ability outside the experimental region is limited. Several examples of such empirical models have been considered in Chapters 3 and 4. Attempting to understand the reasons behind the empirically derived equations can result in the development of empirical mechanistic or mechanistic models.

There are three approaches to set specifications based upon functional relationships: worst-case, statistical, and unified. These are discussed below.

Worst-Case Specifications   These are usually specified as $T \pm \Delta$. T is the target and Δ is the half-width of the specification. The two-sided specification could also be asymmetrical with different values of Δ above and below target. The specification could also be one-sided. By worst-case specifications, we mean that if the input factors are anywhere within their specifications, then the output will be within its specification with 100 percent probability.

Consider a simple example. Two parts A and B are stacked on top of each other to produce stack height y. The functional relationship is

$$y = A + B$$

If the specifications for A and B are 1 ± 0.03, then the worst-case specification for y is 2 ± 0.06. Note that the specification for the output factor is *uniquely* determined based upon the specifications for the input factors.

We are usually interested in the reverse situation, namely, in determining the specifications for input factors given a specification for the output factor. If the specification for y is 2 ± 0.06, what should be the worst-case specifications for A and B? It is easy to see that the specifications for A and B are *not uniquely* determined. There are infinitely many possible specifications. For example, the specification for A and B may be 1 ± 0.03 each, or the specification for A may be 0.5 ± 0.01 and for B may be 1.5 ± 0.05. Specifications for A and B should be selected based upon the degree of difficulty of meeting them in manufacturing such that the P_{pk} index for y is maximized.

In general, if x_i represent the input factors, then for an additive functional relationship

$$y = \alpha_0 + \sum \alpha_i x_i$$

the worst-case specifications (targets and ranges) for y and x_i are related as follows.

$$T_y = \alpha_0 + \sum \alpha_i T_i$$
$$\Delta_y = \sum \alpha_i \Delta_i$$

(5.4)

T represents target and Δ represents the half-width of the specification.

Let us now consider an example of a multiplicative relationship. The drug content of a vial filled with a certain drug substance is a function of fill weight and drug concentration as follows:

$$\text{Drug content} = \text{fill weight} \times \text{drug concentration}$$

The target values are as follows: drug content $= 5$ mg, fill weight $= 50$ mg, and drug concentration $= 0.1$. If the specification for drug content is ± 15 percent of the target, what should be the specifications for fill weight and drug concentration?

Approximately, for multiplicative models, the ± 15 percentage specification on drug content can be allocated as a *percentage specification* to fill weight and drug concentration. This may be understood as follows. Let D denote the drug content, F, the fill weight, and C, the drug concentration. Let $\pm \Delta D$, $\pm \Delta F$, and $\pm \Delta C$ denote the respective ranges. Then

$$D \pm \Delta D = (F \pm \Delta F)(C \pm \Delta C)$$

Neglecting the second-order term

$$D + \Delta D = FC + F\Delta C + C\Delta F$$

Noting that $D = FC$, and then, dividing by D

$$\frac{\Delta D}{D} = \frac{\Delta C}{C} + \frac{\Delta F}{F}$$

This means that the percentage specification for output is the sum of percentage specifications for inputs. Given the specifications for inputs, there is a unique specification for output. The converse is not true. For a ± 15 percentage specification for drug content, there are infinitely many solutions for input specifications, such as a fill weight specification of ± 12 percent of the target and a drug concentration specification of ± 3 percent of the target, or a fill weight specification of ± 5 percent of the target and a drug concentration specification of ± 10 percent of the target, and so on. The actual specifications for inputs should be selected to maximize the P_{pk} index for y.

In general, for multiplicative models

$$y = \alpha_0 x_1^{\alpha_1} x_2^{\alpha_2} \ldots x_k^{\alpha_k}$$

the worst-case specifications (targets and ranges) for y and x_i are related as follows.

$$T_y = \alpha_0 T_1^{\alpha_1} T_2^{\alpha_2} \ldots T_k^{\alpha_k}$$
$$\%\Delta_y = \sum \alpha_i (\%\Delta_i) \tag{5.5}$$

T represents the target and $\%\Delta$ represents the half-width of the specification as a percentage of the target.

For more complex functional relationships, the worst-case specifications for input factors can be determined by Monte Carlo simulation.

Statistical Specifications The worst-case specifications guard against the unlikely events that a characteristic will always be at its specification limit or that all characteristics will simultaneously be at their maximum or minimum limits. If a characteristic is normally distributed with mean centered near the target, characteristic values near the target are far more likely than the values near the edge of the specification. Also, the probability that all characteristics will simultaneously be at their maximum or minimum is extremely low. Statistical specifications account for these probabilities of occurrence.

A statistical specification is denoted by $T \leq \sigma$, where T is the target and σ is the maximum permissible standard deviation. When input factors have statistical specifications, the output factor has a unique statistical specification. As with worst-case specifications, the converse is not true.

Let us now reconsider the previous examples to see how statistical and worst-case approaches differ. Two parts A and B are stacked on top of each other to produce a stack height y. The functional relationship is $y = A + B$. If each part has a statistical specification of $1 \leq 0.01$ (where the 3σ limits correspond to the previously assumed worst-case range of ± 0.03), then uniquely, the specification for the stack height is $2 \leq 0.014$. This is because for additive equations with independent input factors, variances add:

$$\sigma_y^2 = \sigma_A^2 + \sigma_B^2 \tag{5.6}$$

Substituting $\sigma_A = \sigma_B = 0.01$, we get $\sigma_y = 0.014$. Note that the *statistical specification for y is tighter* than the worst-case specification in the sense that the 3σ limits for the statistical specification of y are ± 0.042, whereas the worst-case range for y was ± 0.06.

On the other hand, if the specification for y is $2 \leq 0.02$ (where the 3σ limits correspond to the previously assumed worst-case range of ± 0.06), then from Equation (5.6) one of the infinitely many possible statistical specifications for A

and B is $1 \leq 0.014$. This *specification for input factors is wider* in the sense that the 3σ limits for the statistical specification are ± 0.042, whereas the worst-case range for A and B was ± 0.03.

In general, if x_i represents the input factors, then for an additive model

$$y = \alpha_0 + \sum \alpha_i x_i$$

the statistical specifications (targets and permissible standard deviations) for y and x_i are related as follows.

$$T_y = \alpha_0 + \sum \alpha_i T_i$$

$$\sigma_y = \sqrt{\sum \alpha_i^2 \sigma_i^2}$$

(5.7)

T_i and σ_i represent the target and the individual permissible standard deviations for x_i.

Let us now reconsider the drug content example with the multiplicative model given by

$$\text{Drug content} = \text{fill weight} \times \text{drug concentration}$$

The target values are as follows: drug content $= 5$ mg, fill weight $= 50$ mg, and drug concentration $= 0.1$. If σ for drug content is specified to be ≤ 5 percent of target (corresponding to a 3σ limit of 15 percent), what should be the specifications for fill weight and drug concentration?

For multiplicative equations with independent factors, CV^2 add. This may be understood as follows.

Let D represent the drug content, F, the fill weight, and C, the drug concentration. Then,

$$D = FC$$

and

$$\ln D = \ln F + \ln C$$

Taking derivatives, it follows that approximately

$$\frac{\sigma_D^2}{D^2} = \frac{\sigma_F^2}{F^2} + \frac{\sigma_C^2}{C^2}$$

which leads to the conclusion that

$$CV^2(\text{drug content}) = CV^2(\text{fill weight}) + CV^2(\text{drug concentration})$$

Since the specified percent CV for drug content is 5 percent, $CV^2(\text{drug content})$ 25 (percent)^2 and this can be split up into $CV^2(\text{fill weight})$ and $CV^2(\text{drug concentration})$ based upon the ease of control of the two input factors. For example, if fill weight is more difficult to control than drug concentration, we could have $CV(\text{fill weight}) = 4$

percent and $CV(\text{drug content}) = 3$ percent resulting in a fill weight statistical specification $50 \le 2$ and a drug content statistical specification $0.1 \le 0.003$.

In general, for multiplicative models

$$y = \alpha_0 x_1^{\alpha_1} x_2^{\alpha_2} \ldots x_k^{\alpha_k}$$

the statistical specifications (targets and permissible CVs and standard deviations) for y and x_i are related as follows.

$$T_y = \alpha_0 T_1^{\alpha_1} T_2^{\alpha_2} \ldots T_k^{\alpha_k}$$

$$CV_y = \sqrt{\sum \alpha_i^2 CV_i^2} \tag{5.8}$$

$$\sigma_y = T_y CV_y$$

For more complex functional relationships, the statistical specifications can be determined by Monte Carlo simulation. They should be chosen to maximize the P_{pk} index for y.

Pros and Cons of Worst-Case and Statistical Specifications Worst-case specifications are safe, but overly cautious. They guard against the unlikely events that a characteristic will always be at its specification limit or that all characteristics will simultaneously be at their maximum or minimum specifications. The consequence is that for a given specification on output, worst-case specifications result in tighter (more expensive) specifications on input factors.

Statistical specifications assume the process to be perfectly centered, with a certain standard deviation. Statistical specifications consider the probabilities of various input factors being at various locations. This results in a wider (less expensive) specification on input factors compared with the worst-case specifications. However, the assumption that the process is always perfectly centered is dangerous and off-centered processes can produce defective product even when all inputs are within their statistical specifications. Thus, statistical specifications are overly optimistic and tend to underestimate the range of outputs.

Recall that for the stack height example, the worst-case specifications for A and B was ± 0.03. The statistical specification was wider, in the sense that the 3σ specification limits were ± 0.042. The truth is likely to be somewhere in between. There are many ways in which such a middle path can be found. One approach is to combine the worst-case and the statistical approaches into unified specifications.

Unified Specifications The two approaches can be combined by allowing the process mean to have an operating window defined as a worst-case specification and the individual values to have a certain maximum standard deviation defined as a statistical specification. Such a specification may be written as $T \pm \Delta_m \le \sigma^*$, that is, the process mean has the worst-case specification $T \pm \Delta_m$ and the individual values have a statistical specification with a maximum standard deviation $\le \sigma^*$. Formulae for worst-case specification apply to Δ_m and formulae for statistical specification apply to σ^*.

Given a specification for output, the specifications for inputs can then be derived either by statistical calculations or by simulation.

Let us reconsider the stack height problem. The specification for stack height y can be equivalently stated as $2 \pm 0.02 \leq 0.013$ (corresponding to the 3σ limit of 2 ± 0.06 as before when the mean is at its specification limit). This means that the process mean has to be within 1.98–2.02 with the standard deviation of individual values being ≤ 0.013. It follows that one possible specification for A and B is $1 \pm 0.01 \leq 0.009$.

The results of the three approaches to setting the stack height specification are summarized in the following table.

Approach	Specification for y	Specifications for A and B	Specifications for A and B in 3σ terms
Worst-case	2 ± 0.06	1 ± 0.03	
Statistical	$2 \leq 0.02$	$1 \leq 0.014$	1 ± 0.042
Unified	$2 \pm 0.02 \leq 0.013$	$1 \pm 0.01 \leq 0.009$	1 ± 0.038

As shown in the table, this unified approach represents a middle ground between the worst-case and the statistical specifications. The unified approach is safer than the statistical approach because it does not unrealistically require all inputs to be centered on their targets. It is less expensive than the worst-case approach because it properly considers the probabilities of occurrence. The worst-case and statistical approaches are special cases of this unified approach. When $\sigma^* = 0$, the unified approach becomes the worst-case approach. When $\Delta = 0$, it becomes the statistical approach.

Selecting the Type of Functional Specification

Factors that cause output to vary need to have specifications. Output varies due to three types of factors: external factors (e.g., customer usage factors), product degradation factors, and factors that cause manufacturing variability. Careful consideration is required to select the type of specification for each of these factors.

Ambient temperature and humidity, if they influence the output of a machine, are examples of external factors. For the producer of pizza, bake time and temperature are external customer usage factors. If the pizza is intended to be properly baked over an oven temperature range of 380–420 °F, then this becomes the worst-case specification for oven temperature because an oven may be used at one of these extreme temperatures.

Asymmetric unified specification may be used to account for product deterioration. For example, the drug content of the drug substance in a vial may have a target of 5 mg. The permissible vial-to-vial standard deviation of drug content may be 0.1 mg. Allowing for (say) an operating window of $1.5\sigma = 0.15$ for the mean, the manufacturing specification can be represented by $5 \pm 0.15 \leq 0.1$. If the permissible drug content deterioration over an 18-month shelf life is 0.2 mg, then the specification for drug content will be $5^{+0.15}_{-0.35} \leq 0.1$.

For factors causing manufacturing variation, if the process mean for that factor is specifically controlled to be on target, then a statistical specification may be used.

Otherwise, unified specifications should be used to include an operating window for the mean. If the process can run for extended periods of time far away from the target, then a worst-case specification may be necessary.

5.4   MINIMUM LIFE CYCLE COST APPROACH

It was suggested earlier that there are two ways to view specifications. One approach is the producer's short-term point of view, namely, if the product is within specifications it is good, otherwise it is bad. This leads to the scrap and salvage economic loss function. The economic loss to the producer is a function of the percent of products outside the specification, which have to be scraped or salvaged. The focus is to minimize percent defective. Losses to the customer are not explicitly considered. This is the thinking pattern used in defining the worst-case and statistical specifications.

The Quadratic Loss Function   The second approach is to recognize that an on-target product is the best and deviations from target, even inside the specification, lead to economic losses for the customer. The focus shifts to producing on-target products with specifications set to minimize the total cost to the customer. This total cost includes the purchase price and the losses incurred by the customer throughout the expected life of the product.

The economic loss function selected to quantify customer loss is shown in Figure 5.3 and was described in Chapter 4 as follows. When the product is on target *T*, the average economic loss to the customer is assumed to be zero without loss of generality. If the product is near the target, the economic loss is small. As the deviation increases, the loss increases *more than proportionately*. When the product characteristic has a value *y* immediately after manufacture, the lifetime economic loss averaged over all customers $L(y)$ is given by the following quadratic loss

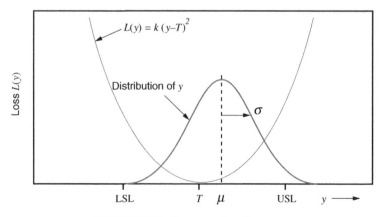

FIGURE 5.3   Economic loss function.

function. The proportionality constant k is estimated by defining a single point on the loss function.

$$L(y) = k(y-T)^2 \qquad (5.9)$$

If the product characteristic immediately after manufacture is distributed normally with mean μ and standard deviation σ (Figure 5.3), then the average loss sustained by the customer can be calculated by integrating the loss function over the distribution.

$$\text{Average loss} = k\left[(\mu-T)^2 + \sigma^2\right] \qquad (5.10)$$

The economic loss to the customer can be reduced by keeping the mean on target and by reducing the variance. Note that when the mean is on target, the economic loss is directly proportional to the variance. As variance is reduced further the economic loss continues to reduce even inside the specification. The use of this quadratic loss function to define signal-to-noise ratios was explained in Chapter 4. Its use to make process improvement and capital investment decisions is explored in Chapter 8.

Setting Specifications Using Quadratic Loss Function   Let us now consider how the specifications can be set using the quadratic loss function. The essential point is to recognize that a producer may be able to take actions to control the key characteristics within any specified limits. Sometimes these actions will cost the producer much less than what they might save the customer, and sometimes they may not. To minimize the total life cycle cost to the customer, specifications can be set so that the cost of control or corrective actions on the part of the producer equals the benefit of these actions to the customer. This thinking pattern shows that the usual specification for drug content of a tablet may be ±15 percent of the target, but if the producer's cost to control the drug content to ±5 percent of the target is less than the economic benefit of this tighter control to the customer, then the specifications for drug content should be ±5 percent of the target.

Consider the example of a voltage converter (Taguchi et al., 1989) where the output voltage may not be on target due to manufacturing variation, product degradation, and environmental factors. A difference of ±20 volts from the target causes the converter not to function properly leading to a customer loss of $100. Each converter is inspected before shipping and the manufacturer can adjust the voltage by adjusting a component at a cost of $5. What should the producer's voltage specifications be?

Figure 5.4 shows this situation. Let y denote the voltage at the time of manufacture and $L(y)$ be the corresponding lifetime loss sustained by the customer. $T\pm\Delta_0$ is the functioning range of the product and A_0 denotes the customer loss at the edge of this range. The producer's specification is represented by $T\pm\Delta$. The producer's cost of

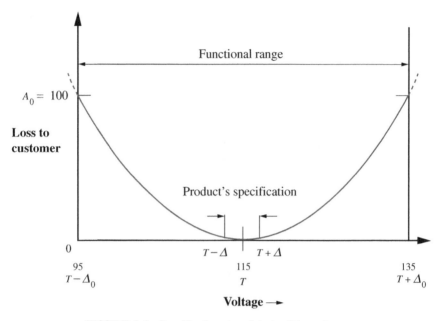

FIGURE 5.4   Specifications to minimize life cycle cost.

adjusting the voltage is A. The loss function is given by

$$L(y) = k(y-T)^2$$

and for a deviation of Δ_0, the loss is A_0. Hence, $k = A_0/\Delta_0^2$.

The manufacturer should adjust the voltage when the customer loss exceeds the cost of adjustment. Not to do so will mean that the customer will lose more money than the producer will save. This is worse than being a thief because a thief plays a zero sum game. Clearly, the producer will be more inclined to do the right thing if the cost of adjustment could be recovered in the short or long term. Customer loss will exceed the cost of adjustment at a value Δ when

$$\frac{A_0}{\Delta_0^2}\Delta^2 \geq A$$

Hence,

$$\Delta = \Delta_0\sqrt{\frac{A}{A_0}} \tag{5.11}$$

In this example, $\Delta_0 = 20$ volts, $A = 5$, and $A_0 = 100$. This gives $T \pm \Delta = T \pm 4.5$ volts as the producer's specification.

Specifications for Multiple Input Factors We now consider the case when multiple input characteristics affect the output, or when there are multiple components that influence the output characteristic of the final product. Let us assume that near the target, the input–output relationship could be approximated by a simple additive linear relationship $y = \alpha_0 + \sum \alpha_i(x_i - T_i)$ or $y = \alpha_0 + \sum \alpha_i|x_i - T_i|$, where $\alpha_0 = T_y$ is the nominal or the target value for y and T_i is the nominal or target value for x_i. If there is a single component or a single factor that controls output, then $y = \alpha_0 + \alpha_1(x_1 - T_1)$ or $y = \alpha_0 + \alpha_1|x_1 - T_1|$. The loss function is

$$L(y) = \frac{A_0}{\Delta_0^2}(y - T_y)^2 = \frac{A_0}{\Delta_0^2}\alpha_1^2(x_1 - T_1)^2$$

where T_y is the target for y and T_1 is the corresponding target for x_1. Let A_1 be the producer's cost of adjusting x_1 or the price of replacing a component. Let the specification for x_1 be $T_1 \pm \Delta_1$. Then Δ_1 should be such that A_1 is equal to the customer's loss. Hence,

$$\frac{A_0}{\Delta_0^2}\alpha_1^2(\Delta_1)^2 = A_1$$

and the optimum half-width of specification for x_1 is

$$\Delta_1 = \frac{\Delta_0}{\alpha_1}\sqrt{\frac{A_1}{A_0}}$$

When there are multiple input factors, the specification for each factor is given by the same formula. For factor i

$$\Delta_i = \frac{\Delta_0}{\alpha_i}\sqrt{\frac{A_i}{A_0}} \qquad (5.12)$$

For worst-case specifications, the relationship between specifications for inputs and the specification for output is

$$\Delta_y = \sum \alpha_i \Delta_i = \frac{\Delta_0}{\sqrt{A_0}}\sum \sqrt{A_i} \qquad (5.13)$$

For statistical specifications, the relationship between specifications for inputs and the specification for output is

$$\Delta_y = \sqrt{\sum \alpha_i^2 \Delta_i^2} = \frac{\Delta_0}{\sqrt{A_0}}\sqrt{\sum A_i} \qquad (5.14)$$

From Equation (5.14), if $\sum A_i < A_0$, then $\Delta_y < \Delta_0$. If y is outside the specification, it is likely in this case that the cost of scrapping the system is smaller than the repair cost and the system may be scrapped. If $\sum A_i > A_0$, then $\Delta_y > \Delta_0$, that is, the specification for y exceeds the functional limit. In this case, the cost of repair is hopefully much less than scrapping the system such that the system could be salvaged. When $\sum A_i = A_0$, then $\Delta_y = \Delta_0$ and if y is outside the specification limits, it would be scrapped or salvaged depending upon the cost.

Let us consider an example. A product consists of two components x_1 and x_2. The component prices are $A_1 = \$100$ and $A_2 = \$200$, respectively. Each component characteristic has an approximately linear effect on the product's characteristic y around the nominal values of the components with coefficients $\alpha_1 = 2$ and $\alpha_2 = 4$. The relationship is given by

$$y = 10 + 2x_1 + 4x_2$$

The functional tolerance for the product characteristic is ± 4 or $\Delta_0 = 4$ and the loss at this functional limit is $A_0 = \$1000$. What is the specification for each component?

If Δ_1 and Δ_2 represent the half-widths of the specifications for x_1 and x_2, then from Equation (5.12),

$$\Delta_1 = \frac{4}{2}\sqrt{\frac{100}{1000}} = 0.63$$

$$\Delta_2 = \frac{4}{4}\sqrt{\frac{200}{1000}} = 0.44$$

If $\pm \Delta_y$ represents the specification for y, then from Equations (5.13) and (5.14),

$$\Delta_y(\text{worst case}) = \frac{4}{\sqrt{1000}}\left(\sqrt{100} + \sqrt{200}\right) = 3.1$$

$$\Delta_y(\text{statistical}) = \frac{4}{\sqrt{1000}}\left(\sqrt{100 + 200}\right) = 2.2$$

Summary   Specifications are written as follows: worst case: $T \pm \Delta$, statistical: $T \leq \sigma$, and unified: $T \pm \Delta_m \leq \sigma^*$. Linear additive and multiplicative models and targets are given by:

Linear additive model: $y = \alpha_0 + \sum \alpha_i x_i$ and $T_y = \alpha_0 + \sum \alpha_i T_i$

Multiplicative model: $y = \alpha_0 x_1^{\alpha_1} x_2^{\alpha_2} \ldots x_k^{\alpha_k}$ and $T_y = \alpha_0 T_1^{\alpha_1} T_2^{\alpha_2} \ldots T_k^{\alpha_k}$

Table 5.1 summarizes the various formulae used to set specifications for additive and multiplicative models. The targets and permissible ranges should be jointly set to permit wide ranges for input factors while achieving the desired P_{pk} for y. More complex models need Monte Carlo simulation solutions.

TABLE 5.1   Summary of Specifications

Specification Type	No Model	Additive Model	Multiplicative Model
Empirical	$\bar{x} \pm ks$		
Worst case		$\Delta_y = \sum \alpha_i \Delta_i$	$\%\Delta_y = \sum \alpha_i(\%\Delta_i)$
Statistical		$\sigma_y = \sqrt{\sum \alpha_i^2 \sigma_i^2}$	$\sigma_y = T_y\sqrt{\sum \alpha_i^2 CV_i^2}$
Unified		$\Delta_{my} = \sum \alpha_i \Delta_{mi}$	$\%\Delta_{my} = \sum \alpha_i \%\Delta_{mi}$
		$\sigma_y^* = \sqrt{\sum \alpha_i^2 \sigma_i^{*2}}$	$\sigma_y^* = T_y\sqrt{\sum \alpha_i^2 CV_i^2}$
Minimize total cost (worst case)		$\Delta_i = \dfrac{\Delta_0}{\alpha_i}\sqrt{\dfrac{A_i}{A_0}}$	
		$\Delta_y = \dfrac{\Delta_0}{\sqrt{A_0}}\sum \sqrt{A_i}$	
Minimize total cost (statistical)		$\Delta_i = \dfrac{\Delta_0}{\alpha_i}\sqrt{\dfrac{A_i}{A_0}}$	
		$\Delta_y = \dfrac{\Delta_0}{\sqrt{A_0}}\sqrt{\sum A_i}$	

5.5   QUESTIONS TO ASK

The following are the type of questions to ask to help improve the understanding and the setting of specifications.

1. How are the specifications being currently developed in your company? Does the process of setting specifications need to be improved?
2. What are the R&D and manufacturing implications of specifications mandated by the customer? Have you set appropriate variability targets for R&D and manufacturing necessary to meet the specifications?
3. Have the key characteristics that control product performance been identified?
4. Are the targets for key characteristics correctly set? What methods were used to set the targets?
5. Which approach is appropriate to set the specifications for the project at hand—empirical, functional, or life cycle cost approach?
6. If an empirical approach to setting specifications is being used, are the half-widths of specifications being set as two or three times the observed standard deviation? If so, the tolerance interval method described here coupled with scientific judgment should be used.
7. Are attempts being made to get a better understanding of functional relationships between the input and output factors? How are these functional relationships being determined? How predictive are the relationships?

8. If a functional approach to setting specifications is being used, is the type of specification (worst case, statistical, or unified) correctly selected for each key characteristic?

9. Are attempts being made to understand the cost structure in terms of the cost of controlling key factors or replacing key components?

10. Have attempts been made to understand and apply the life cycle cost approach? If not, a pilot project should be started to make simpler applications and compare the results to other approaches currently being taken.

Further Reading   The book by Joglekar (2003) discusses the implications of variance component analysis on setting specifications. It also includes a discussion of how the specification may be allocated to R&D (product design) and to manufacturing (process control). Both single level and multi-level specifications are considered. The book by Taguchi et al. (1989) discusses the life cycle cost approach to setting specifications. It includes an extensive discussion of loss function and the use of life cycle cost approach to setting specifications for single and multiple characteristics when the characteristic types are smaller-the-better, larger-the-better, and nominal-the-best.

CHAPTER 6

HOW TO DESIGN PRACTICAL ACCEPTANCE SAMPLING PLANS AND PROCESS VALIDATION STUDIES?

The design of acceptance sampling plans and process validation studies is often inadequately done in industry. One purpose of this chapter is to clarify some misconceptions that exist in industry regarding the protection provided by acceptance sampling plans. The second purpose is to provide practical guidance to select acceptable quality level (AQL) and rejectable quality level (RQL) necessary to design acceptance sampling plans. Once the values of AQL and RQL are selected, the design of acceptance sampling plans is automatic and can be delegated to a software package. The third purpose is to explain the connection between process validation studies and acceptance sampling plans to properly design process validation studies.

Misconceptions in Industry Acceptance sampling plans are used in manufacturing to accept or reject incoming, in-process, and finished product lots. Process validation studies are used to accept or reject a manufacturing process before its transfer to manufacturing. For the practitioner, the statistical calculations necessary to design acceptance sampling plans and process validation studies are not that important today because the calculations have been converted to standards, are often available as corporate standard operating procedures, and have been codified in software packages. Many of the standards are indexed by AQL, and selecting a value of AQL results in a designed acceptance sampling plan. Those who design plans in this manner are sometimes not even aware that every sampling plan has a RQL as well! The emphasis on understanding the implications of a sampling plan through its operating

Industrial Statistics: Practical Methods and Guidance for Improved Performance   By Anand M. Joglekar
Copyright © 2010 John Wiley & Sons, Inc.

characteristic curve has reduced. This has resulted in misconceptions regarding the protection provided by the sampling plan, along with a lack of explicit consideration of whether that protection is adequate for the specific project at hand. The values of AQL and RQL are at times selected without a thorough consideration of their practical implications. Sometimes people think that consumers should only be concerned about the definition of RQL, and they should have no say in the selection of AQL! Terminology such as consumer's point and producer's point on the operating characteristic curve creates this impression. The role of current quality level (CQL), the quality that can reasonably be produced by the manufacturing process, in assessing the risks of an acceptance sampling plan is often not considered. It is not uncommon for acceptance sampling plans to be designed without knowing CQL. Finally, in designing process validation studies, the connection between the design of process validation studies and acceptance sampling plans to be used for lot-by-lot manufacture has not always been understood resulting in inadequately designed validations.

This chapter deals with three practically important topics:

1. Use of the operating characteristic curve of a sampling plan to clarify some misconceptions that exist in industry.
2. Explanation of the practical considerations necessary to correctly select AQL and RQL. Once AQL and RQL are selected, the design of the sampling plan is automatic and can be delegated to a software package.
3. Proper design of validation studies based upon the connection between process validation studies and acceptance sampling plans to be used in manufacturing.

The R&D and Production Interface   Design of process validation studies and acceptance sampling plans occurs as the project transitions from R&D to manufacturing. The sequence of events is the following:

process characterization $\rightarrow$ process validation $\rightarrow$ lot-by-lot acceptance

sampling $\rightarrow$ process control and improvement

R&D usually has the major responsibility for process characterization and validation, while manufacturing is responsible for lot-by-lot acceptance sampling, process control, and continuous process improvement. The main purposes of these four steps are as follows.

Process Characterization: The purpose is to identify the key factors that govern product performance and to establish specifications for the key factors within which the product performance is satisfactory. Design of experiments is an appropriate tool to use at this stage. Unfortunately, in many instances, this step is either skipped or inadequately performed, resulting in poor process and raw material specifications and expensive fixes in manufacturing.

Process Validation: The purpose is to prove the suitability of transferring the process to manufacturing by demonstrating the ability to produce products with sufficiently low fraction defectives so that the probability of lot rejection in manufacturing will be acceptably small.

Lot-by-Lot Acceptance Sampling: The purpose is to prevent the shipment of an occasionally produced poor quality (high fraction defective) lot by testing a sample of products from each lot and either accepting or rejecting each lot.

Process Control and Improvement: One purpose is to control the manufacturing process by identifying and correcting process shifts as they occur. A second purpose is to continuously improve product quality and reduce cycle times and cost. Statistical process control and variance components analysis are among the useful tools at this stage.

Let us now turn to the topics of interest in this chapter, first, acceptance sampling plans and, next, process validation studies, because the design of process validation studies should be based upon acceptance sampling plan concepts.

The purpose of a sampling plan is to decide, based upon a representative sample, whether to accept or reject a lot. Each inspected item in the sample may be classified as either acceptable or not acceptable. This is attribute data collection and the sampling plans are known as attribute sampling plans. There are variable sampling plans where each inspected item is measured with respect to some characteristic on a continuous scale. For attribute or variable data, the decision to accept or reject a lot can be made on the basis of a single sample, two samples, or multiple samples taken from a lot. The resultant sampling plans are known as single, double, or multiple sampling plans.

6.1   SINGLE-SAMPLE ATTRIBUTE PLANS

The practical considerations necessary to correctly design and interpret the various types of sampling plans can be explained in the context of a single-sample attribute sampling plan as follows.

The Framework   Consider a lot consisting of some number of discrete items. An accept/reject decision is to be made with respect to this lot. There are two parties to this decision, the producer and the consumer. The lot should be accepted if it is good and rejected if it is bad. Exactly what do we mean by a good lot? If we mean a lot that contains no defectives, then it will require 100 percent inspection to weed out any possible defective item. This may be very costly and is impractical for destructive tests. If 100 percent inspection is impractical, we could evaluate a small sample of products from the lot and make our accept/reject decision based upon the small sample results. This is what an acceptance sampling plan does. However, on the basis of a sample, there is no way to guarantee that the lot contains no defectives.

This question of defining a good lot was front and center in the early discussion and development of acceptance sampling plans in the 1940s for the department of defense. The recognition that sampling plans could not be developed to accept only perfect lots meant that good lots had to be defined as containing some fraction defective, however small. This led to the definition of acceptable quality level (AQL).

AQL = the largest percent defective that is acceptable as a process average

The following comments made by H. R. Bellinson (paraphrased here for brevity) regarding the difficulty of introducing the AQL concept in 1942, reported at the Acceptance Sampling Symposium, American Statistical Association, Washington, D.C., 1950 are instructive (Grant and Leavenworth, 1980).

> A unique feature of the acceptance sampling plans was the concept of AQL. This concept had to be introduced because there was no way to develop a sampling plan for attributes on the basis of accepting only perfect material. We had to say that we will accept a certain fraction defective, make that as small as you please. That may sound as though we proposed to degrade quality, but the fact was that the fraction defective being accepted under the methods then in use was much larger than anything we proposed.
>
> There were difficulties selling the concept. The engineering department was concerned about accepting any defectives whatever. But then they thought that if the fraction defective was small, the tail of the normal distribution could not go too far and the factor of safety may take care of it.
>
> The fiscal department had a different point of view. It was that the government had a contract to pay for 100% perfect material and they did not want to pay for the 1% defectives. After we were able to prove that the proposed sampling plan was much more economical than the former methods of inspection, they gave us their blessings.
>
> The legal department had yet another point of view. They were not concerned about accepting a few bad ones; they were concerned with the fact that we were going to reject many good ones. If we reject a lot that is 2% defective the lot is still 98% perfect. The legal department said that we had no right to reject individual pieces that were perfect. The argument was settled by changing requirements.

Suppose the acceptable quality level (AQL) = 1 percent. This means that a lot containing 1 percent defective is being classified as a good lot. It may be tempting to say that a bad lot will be one with greater than 1 percent defectives. However, it is easy to see that without 100 percent inspection, we cannot distinguish between a lot containing a shade less than 1 percent defectives and another with a shade more than 1 percent defectives. This leads to the definition of bad lots with RQL > AQL.

RQL = the largest percent defective at which an individual lot should be rejected

If AQL = 1 percent and RQL = 5 percent, it would mean that we expect the process average percent defective to be 1 percent or less and we have to reject any individual lots with more than 5 percent defectives.

The acceptance sampling plan is designed such that lots with percent defectives $\leq$ AQL will be accepted most of the time and lots with percent defectives $\geq$ RQL will be rejected most of the time.

Operating Characteristic Curve   An attribute sampling plan generally takes the following form. Select a representative sample of n products from the lot and classify each as defective or not defective. If the number of defectives $x \leq c$, accept the lot, otherwise reject. The sample size n and the acceptance number c are the two parameters of the sampling plan.

There are several ways in which a representative sample may be selected. One approach is random sampling, where each product in the lot has an equal chance of being selected. Another approach is periodic sampling, where a product is selected from the manufacturing line at a periodic time interval dictated by the sample size. A third approach is stratified sampling, where the lot is considered to be subdivided on some rational basis and a sample is taken from each subdivision in proportion to the size of that subdivision. For example, if there are four workstations on a machine, then 25 percent of the required samples could be randomly selected from each workstation.

Every sampling plan has an operating characteristic curve associated with it. Consider a plan with $n = 50$ and $c = 0$. This means that the lot will be accepted if no defectives are found in a sample of size 50. If p denotes the true fraction defective in the lot, then the probability of lot acceptance P_a is given by

$$P_a = (1-p)^{50}$$

As the true fraction defective p increases, the probability of lot acceptance reduces. Figure 6.1 ($n = 50$ and $c = 0$) shows this relationship with true percent defective ($100p$)

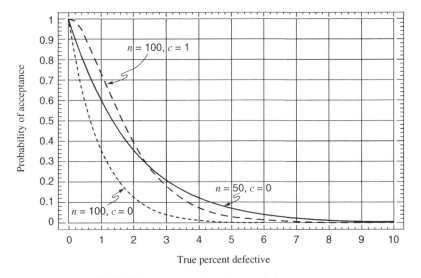

FIGURE 6.1   Operating characteristics curves.

on the x-axis and probability of lot acceptance on the y-axis. This curve is called the operating characteristic curve for the sampling plan. When the lot has zero percent defective, the probability of lot acceptance is 100 percent. When the lot is 100 percent defective, the probability of lot acceptance is zero percent. This operating characteristic curve shows that a lot with 1 percent defectives will be accepted about 60 percent of the time, a lot with 5 percent defectives will be accepted about 10 percent of the time, and so on. If AQL = 1 percent and RQL = 5 percent, good lots will be rejected 40 percent of the time and bad lots will be accepted 10 percent of the time by this sampling plan. The operating characteristic curve is the best way to understand the consequences of any acceptance sampling plan in terms of the probabilities of accepting or rejecting lots of specified quality, and thus, the protection provided to the consumer and the producer.

The shape of the operating characteristic curve depends upon the values of n and c. As shown in Figure 6.1 ($n = 100$ and $c = 0$), as n increases, the curve becomes steeper. As c increases ($n = 100$ and $c = 1$), the curve becomes flatter at the top. For AQL = 1 percent and RQL = 5 percent, with $n = 100$ and $c = 1$, the probability of rejecting a good lot reduces to about 25 percent and the probability of accepting a bad lot reduces to about 5 percent.

Given the shape of the operating characteristic curve, it becomes clear that for any value of n and c, good lots with a percent defective equal to AQL will always be rejected with some probability α and bad lots with a percent defective equal to RQL will always be accepted with some probability β. These are known as the producer's and the consumer's risks, respectively, defined in Equation (6.1) and shown in Figure 6.2.

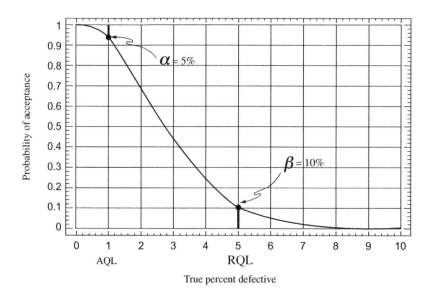

FIGURE 6.2   Designed sampling plan (AQL = 1 percent, $\alpha = 5$ percent), (RQL = 5 percent, $\beta = 10$ percent), ($n = 132$, $c = 3$).

Producer's risk (α) = Probability (rejecting a lot with percent defective = AQL)

Consumer's risk (β) = Probability (accepting a lot with percent defective = RQL)

$$(6.1)$$

Designing an Acceptance Sampling Plan   Designing an acceptance sampling plan means finding the values of n and c that produce the desired operating characteristic curve. The desired operating characteristic curve is defined by specifying the two points on the curve, sometimes called the producer's point (AQL, α) and the consumer's point (RQL, β). It has become a common practice to have $\alpha = 5$ percent and $\beta = 10$ percent, although with software packages, any desired values can be specified for these two risks. Once these two points on the operating characteristic curve are specified, the values of n and c can be obtained such that the curve nearly goes through these two points.

For example, suppose the two points on the operating characteristic curve are (AQL = 1 percent, $\alpha = 5$ percent) and (RQL = 5 percent, $\beta = 10$ percent) (Figure 6.2). The values of n and c are obtained by solving the producer's risk and consumer's risk Equation (6.1). Understanding the exact statistical equations and their solution (Grant and Leavenworth, 1980) used to be important before the advent of software packages. The job can now be delegated to a computer program (or appropriate standards) and the designed sampling plan has $n = 132$ and $c = 3$. As shown in Figure 6.2, the operating characteristic curve goes through the designated points almost exactly. Thus, the design of the sampling plan is essentially complete when AQL and RQL are selected.

Deal or No Deal?   The purpose of the following discussion is to ensure that the implications of the operating characteristic curve are well understood and to clear up some of the misconceptions that exist in industry.

1. *An Accepted Lot Is a Good Lot:* This statement in not necessarily true. From Figure 6.2, it is easily seen that if a lot contains 4 percent defectives, it will be accepted 25 percent of the time. Even a lot containing 6 percent defectives will be accepted 5 percent of the time. Clearly, just because a lot is accepted by a sampling plan does not mean that it is a good lot. What can we say about the quality of an accepted lot? We can say that if a lot is accepted, we are $100(1 - \beta)$ percent sure that the true fraction defective p in that lot is $\leq$RQL. For RQL = 5 percent and $\beta = 10$ percent, we are 90 percent sure that the accepted lot has a fraction defective $p \leq 5$ percent.

2. *A Rejected Lot Is a Bad Lot:* This statement is also not necessarily true. From Figure 6.2, it is easily seen that a good lot with 1 percent defectives will be rejected 5 percent of the time. What we can say about a rejected lot is that we are $100(1 - \alpha)$ percent sure that the rejected lot has a fraction defective $p \geq$ AQL. For AQL = 1 percent and $\alpha = 5$ percent, we are 95 percent sure that the rejected lot has a fraction defective $p \geq 1$ percent.

3. *Acceptance Sampling Is Counterproductive If the Manufacturing Process Is Perfectly Stable:* This is a true statement. A perfectly stable process (see Chapter 7) means that the true fraction defective in each lot is the same. Suppose it is equal to 2 percent. Then, if we apply the acceptance sampling plan in Figure 6.2 to such a stable process, approximately 72 percent of the lots will be accepted and 28 percent of the lots will be rejected. However, all the lots have the same percent defectives, either all of them should be rejected or none of them should be rejected. The application of an acceptance sampling plan to accept or reject the same quality lots is counterproductive.

4. *Acceptance Sampling Helps Weed Out Occasional Production of Bad Lots:* This is a true statement. When percent defectives vary from lot to lot, acceptance sampling plans can be the most economical choice compared with alternatives such as releasing all lots without testing or 100 percent inspecting all lots. Acceptance sampling plans assume that bad lots will be occasionally produced and will be rejected by the sampling plan most of the time.

5. *We Use an Accept-on-Zero Plan (c = 0) Because the Plan Guarantees That All Accepted Lots Have No Defectives in Them:* This statement is not true. Just because the sample did not contain a defective is no guarantee that the lot does not contain a defective. Accept-on-zero plans have the smallest sample size. They have an exponentially falling operating characteristic curve (see Figure 6.1). This usually implies larger producer's risk and does put greater pressure on the producer to improve quality. If the process average percent defective is significantly below the AQL, the accept-on-zero plan may be economical for the producer as well.

6. *As AQL and RQL Get Closer Together, the Sample Size Increases:* This statement is true. If AQL and RQL are closer together, the operating characteristic curve will be steeper, and steeper curve means larger sample size (see Figure 6.1).

7. *Sample Size Is a Function of the Lot Size:* This statement is partially true. For the plans given in ANSI standards, the sample size increases with the lot size because it is felt that a larger sample size is necessary to obtain a representative sample from a large lot and not because of statistical considerations. In general, the sample size is essentially not a function of the lot size so long as the sample represents the population. The exception is that if the sample size becomes 10 percent or more of the lot size, then the lot size does have some influence on the sample size. In most practical situations, the sample size is considerably smaller than 10 percent of the lot size. The fact that the sample size is not a function of the lot size means that the cost of sampling can be reduced by increasing the lot size. If so, what is the incentive to keep the lot size small? The key assumption is that the lot is homogeneous, namely, portions of manufacturing with completely different fraction defectives have not been commingled. Large lots may make homogeneity less likely.

8. *Consumer Should Decide the Consumer's Point (RQL, β) and Producer Should Decide the Producer's Point (AQL, α) on the Operating Characteristic*

Curve: This is not a true statement. The consumer does seek protection against accepting poor quality represented by RQL and the producer wants protection against rejecting good quality represented by AQL. However, the consumer is interested in defining what the AQL should be because that is the quality the consumer is likely to get most of the time. In fact, the consumer is at least as much interested in defining AQL as in defining RQL. On the other hand, if the AQL is arbitrarily defined to be too low, many reasonably good lots may get rejected and the cost of such rejection is likely to be passed on to the consumer. Similarly, if RQL is defined to be very close to AQL, not only will the cost of sampling go up substantially, but also the lots very close to AQL quality will be rejected. Thus, consumers and producers should not be viewed as adversaries in the selection of AQL and RQL, with producers wanting to move both to the right and consumers wanting to move both to the left. This can be understood better by realizing that many a time, both the producer and the consumer belong to the same company (internal supplier–customer relationships). Decisions regarding selection of AQL and RQL need to be more collaborative than adversarial, with the interests of both parties considered in deciding both the AQL and the RQL, even when the decisions are made solely by the producer.

6.2 SELECTING AQL AND RQL

The practical question in designing acceptance sampling plans has to do with the selection of AQL and RQL. Once these are selected, the design of the plan is automatic. The factors to consider in selecting AQL and RQL are discussed below.

Criticality   This refers to the consequences of an unacceptable or out-of-specification product. In attribute testing, the product is classified as unacceptable due to certain defects. In variable testing, the product is unacceptable if the characteristic being measured is out of specification. The greater the harmful consequences of an unacceptable product, the smaller the AQL and RQL will have to be. These harmful consequences are usually assessed by conducting a risk assessment to determine the severity of the consequence and the probability of the occurrence of the consequence. The risk assessment leads to classification of characteristics by the degree of harmful effects of an unacceptable product. Critical, major, and minor are the commonly used classifications of characteristics and these terms are defined as follows:

1. *Critical:* Unacceptable product poses safety concerns or may ruin company finances.
2. *Major:* Unacceptable product substantially degrades product function or may substantially negatively impact company finances.
3. *Minor:* Unacceptable product has little effect on product function or company finances.

Sometimes additional classifications are used. The values of AQL and RQL are selected to reflect the harmful consequences. For example, the percent AQL and RQL used by one company are as follows: critical (0.1, 0.5), major A (0.25, 1.0), major B (0.65, 2.5), minor A (1.0, 6.0), and minor B (2.0, 10.0). For critical characteristics, this company states that a lot containing 1 defective in a 1000 items is acceptable, but a lot containing 5 defectives in a 1000 items is not acceptable.

Industry Practice Certain industry practices have evolved over the years and these provide useful guidance for selecting AQL and RQL. For example, in one industry, the five values of AQL used in increasing order of criticality are 10, 5, 3, 1, and 0.1 percent. In some industries, the values are contractually specified, as in the case of the Department of Defense (DoD). Also, the Food and Drug Administration (FDA) has sometimes required the use of certain specified sampling plans, with implied AQL and RQL.

As an example of a mandated sampling plan, the level 1 acceptance sampling plan required by FDA for tablet drug content has $n = 10$ and $c = 0$. Figure 6.3 shows the operating characteristic curve corresponding to this plan. The AQL is approximately 0.25 percent matching the major A classification discussed above. The RQL is 20 percent. A lot containing 7 percent out-of-specification tablets will be accepted approximately 50 percent of the time by this plan. Clearly, the plan by itself does not provide sufficient protection to the consumer against an occasional production of bad lot.

On the other hand, if the true average percent defective is greater than 0.25 percent, a large number of lots will be rejected and this is a great motivation for the producer to have adequate engineering controls to ensure that the tablet drug content is within specifications. This plan may be reasonable for a producer with a

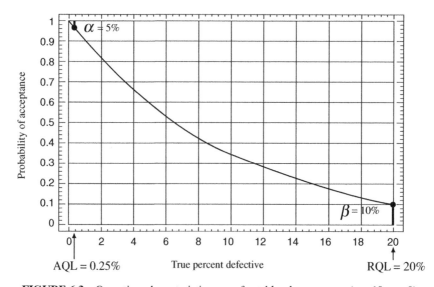

FIGURE 6.3 Operating characteristic curve for tablet drug content ($n = 10$, $c = 0$).

history of good production. It does not sufficiently protect the consumer against an occasionally produced bad lot. A partial solution to this problem is provided by switching schemes.

Switching Schemes   The often-used ANSI standard for the design of acceptance sampling plans is intended to be applied as an acceptance sampling *scheme*, involving the use of normal, tightened, and reduced inspection. There are rules for switching between the three inspection levels intended to penalize the producer when quality degrades, and reward the producer for good quality. For example, when normal inspection is in effect, tightened inspection is instituted when two out of five consecutive lots are rejected. Tightened inspection effectively reduces both the AQL and the RQL, providing much greater protection to the consumer when there is indication of poor quality. A common abuse of the standards is the failure to use switching rules, resulting in a substantial increase in consumer's risk.

A more extreme version of switching may be explained as follows. Acceptance sampling plan is a hypothesis test. When we trust the producer, the null hypothesis is $H_0: p \leq$ AQL. In other words, we choose to believe that the quality is good and require evidence to reject the hypothesis to conclude that quality is unsatisfactory. If we did not trust the producer, as may be the case when lots fail or other indications of poor quality exist, then the hypothesis may be reversed, $H_0: p \geq$ AQL, implying that we believe the quality to be poor and need evidence to accept that it is good. In an extreme case, this will put the producer in a position of having to revalidate the process, as explained later under process validation.

Collaborative Selection of AQL   As suggested earlier, consumers and producers need not be considered as adversaries and this is indeed the case with longstanding customer–supplier relationships and when both parties belong to the same parent company as internal customers and suppliers. The AQL could then be selected to minimize the total cost.

Figure 6.4 shows the likely cost structure for the producer and the customer as a function of true fraction defective p. As the fraction defective in purchased material increases, customer's cost associated with having to deal with bad product increases. Similarly, the producer's costs are likely to be higher to produce products with very low fraction defective. The sum of the producer's and customer's costs will be a parabolic function of p as shown in Figure 6.4. The value of p that minimizes total cost should be used as AQL.

The x-Axis Must Spell CAR   The current quality level (CQL) is the true mean fraction defective produced by the manufacturing process. To prevent excessive lot rejections in manufacturing, it is important to ensure that CQL $\leq$ AQL $<$ RQL. As shown in Figure 6.5, the x-axis must spell CAR. For example, if CQL is between AQL and RQL, a large proportion of normal production will be rejected. If the x-axis was to spell ARC, it would indeed be a fiery end to manufacturing!

Often in industry, AQL is selected without understanding, or paying attention to, what the CQL is. This is a big mistake.

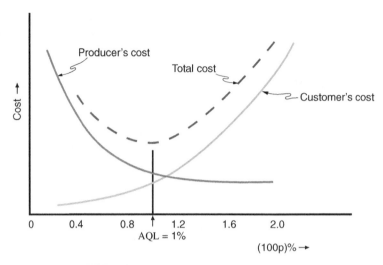

FIGURE 6.4   Selecting AQL to minimize total cost.

Estimating CQL   How can we estimate the current quality level before designing acceptance sampling plans? The CQL can be estimated based upon data from prior engineering studies, characterization studies, and validation studies. The calculations for attribute data are shown below. The statistical basis for the calculations and the calculations for variable data are explained in Joglekar (2003).

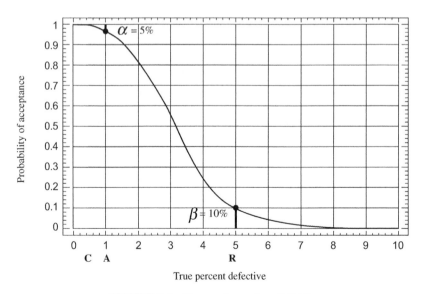

FIGURE 6.5   The *x*-axis must spell CAR.

Suppose that prior data show one defective product in 100 tested. This means that the estimated fraction defective is 0.01. The true fraction defective p may be different from this estimate. The value of p_{max}, the upper confidence limit for p, is given by

$$p_{max} = \frac{\lambda}{n}$$

where the values of λ are tabulated below corresponding to the number of defectives x found in the n examined items.

Percent confidence	$x=0$	$x=1$	$x=2$	$x=3$	$x=4$
90%	2.30	3.89	5.32	6.68	7.99
95%	3.00	4.74	6.30	7.75	9.15

As examples, for $x=1$ and $n=100$, we are 90 percent sure that $p_{max} = 3.89/100 = 0.0389$ or 3.89 percent. For $x=0$ and $n=200$, we are 95 percent sure that $p_{max} = 3.00/200 = 0.015$ or 1.5 percent. These calculations should be done before selecting the AQL.

Note that the formula can be used in reverse to calculate the sample size necessary to demonstrate that p is less than any specific value. The smallest necessary sample size will correspond to $x=0$. To show with 95 percent confidence that $p \leq 0.01$, $n = 3.00/0.01 = 300$. If no defectives are found in a sample of 300 products, then we will be 95 percent sure that the true fraction defective is less than or equal to 1 percent or the CQL ≤ 1 percent. The purpose of validation studies is to demonstrate that the CQL is acceptable.

6.3   OTHER ACCEPTANCE SAMPLING PLANS

So far we have considered single sample attribute plans. There are many other types of attribute sampling plans including double, multiple, and sequential sampling plans. Similar sampling plans exist for variable characteristics measured on a continuous scale. In comparing these various plans to each other, it is most important that plans be compared with each other in terms of their relative advantages and disadvantages under the condition that they have the same operating characteristic curve. Otherwise, the risks associated with the plans being compared will be different and the comparisons will be invalid. The key considerations in selecting AQL, RQL, and the two risks remain the same as described previously. Once the values of AQL and RQL are specified, the actual design of these various types of sampling plans is delegated to a computer program, or tables of designed plans such as those in the military standards or ANSI standards are used. We now briefly describe the more commonly used of these sampling plans and their pros and cons. For a detailed discussion of these various sampling plans, the reader is referred to Grant and Leavenworth (1980).

Double Sampling Plans   In a double sampling plan, under certain circumstances, a second sample is taken before making an accept/reject decision. There are four parameters:

n_1 = sample size for the first sample
c_1 = acceptance number for the first sample
n_2 = sample size for the second sample
c_2 = acceptance number for the total sample

For example, $n_1 = 40$, $c_1 = 1$, $n_2 = 100$, and $c_2 = 2$. If the number of defectives in the first sample of size 40 is ≤ 1, the lot is accepted. If the number of defectives in the first sample is >2, the lot is rejected. Otherwise, a second random sample of size 100 is taken. If the total number of defectives in the first and second samples is >2, the lot is rejected, otherwise it is accepted.

There are two potential advantages of a double sampling plan. One is psychological, just the idea of having a second chance. The second key advantage is that the expected sample size, and therefore, the cost of testing, for a double sample plan, are very often smaller than those for a single sample plan. This is because the sample size for the first sample of a double sampling plan is smaller than the sample size for a single sample plan with the same operating characteristic curve as the double sample plan. Most of the time, the lot will be accepted or rejected on the first sample. Also, it could be rejected on the second sample without completing the total inspection.

There are two potential disadvantages to a double sampling plan. Under some circumstances, the expected sample size for a double sampling plan could exceed that for a single sampling plan, thus negating the cost benefits. The second disadvantage is that the double sampling plan may be more difficult to administer.

Variable Sampling Plans   Variable sampling plans are used when data are collected on a continuous scale. Variable sampling plans assume a normal distribution for data. A variable sampling plan is characterized by the sample of size n and acceptance constant k and the accept/reject decisions are made based upon sample average and standard deviation. The lot is accepted if $\bar{x} \pm ks$ is within the specification for that characteristic. The value of k is determined depending upon whether the standard deviation is assumed to be known or unknown, and whether the specification is one- or two-sided.

The main advantage of a variable sampling plan is that the same operating characteristic curve is obtained with a much smaller sample size compared with the attribute sampling plan. This reduction in sample size occurs because variable data contain more information than attribute data. While variable measurements are likely to cost more than attribute measurements, the large reduction in sample size often leads to reduced cost of testing. When AQL is very small, the sample sizes for attribute sampling plans are so large that there is no choice other than to use a variable sampling plan.

There are three disadvantages of a variable sampling plan. The plan assumes the data to be normally distributed, and departures from normality cause the fraction defective to be wrongly calculated leading to wrong decisions. More specifically, they require that the tails of the distribution have a probability mass no greater than that for a normal distribution. If the data are not normally distributed, data transformation or other techniques may have to be used. If such suitable actions are not found then an attribute sampling plan is used. The second disadvantage is that a separate sampling plan is required for each key product characteristic. Finally, the plan may occasionally reject a lot even when none of the observations are outside the specification and people may have psychological difficulty with this situation.

ANSI/ASQC Standards Z1.4 and Z1.9 are two commonly used standards to select attribute and variable sampling plans, respectively (ANSI/ASQC, 1981). These standards include single, double, and multiple sampling plans. Provision is made for normal, tightened, and reduced inspection and switching rules are provided to move from one type of inspection to another. The plans are indexed by specified values of AQL and risks, the inspection level (criticality), and the lot size. While lot size has almost no effect on the sample size as explained before, lot size is included here because it is felt that getting a representative sample from a larger lot is more difficult than getting it from a smaller lot. Today, it is possible to tailor the sampling plans to the desired AQL, RQL, and risks by using software packages without having to use tabled values.

6.4 DESIGNING VALIDATION STUDIES

One approach to process validation is to produce a small number of lots, usually from one to three lots, with all in-process controls held at target (sometimes, challenge conditions are used as well) and examine n items per lot. Validation passes if for each lot $x \leq c$, where x represents the number of defectives in each lot and c is the acceptance number. This is exactly the same structure as that for the attribute acceptance sampling plan. Similarly, for variable characteristics a variable sampling plan is used to design validation studies.

What should be the values of AQL and RQL in designing validation studies? If AQL_m and RQL_m represent the values of AQL and RQL to be used in manufacturing, it is tempting to say that the same values should be used in validation. This is in fact how some companies seem to address validation. Here is the guidance provided by one company for process validation.

Criticality	Percent Confidence	Percent Reliability	Attribute Plan	
			n	c
Critical	95	99	298	0
Major	95	95	59	0
Minor	90	90	25	0

Reliability is the percent of acceptable products. According to this company, 95 percent confidence and 95 percent reliability are satisfactory requirements for a major characteristic. This defines the consumer's point on the operating characteristic curve to be $RQL = 5$ percent and β risk $= 5$ percent, which matches RQL_m used by this company in manufacturing. With $n = 59$ and $c = 0$, if validation passes, we would have demonstrated that we are 95 percent sure that the true fraction defective is not greater than $RQL_m = 5$ percent.

However, the purpose of the validation is different. It is to demonstrate that $CQL \leq AQL_m$, namely, the true quality level of the manufacturing process is better than the acceptable quality level in manufacturing (AQL_m) (not just better than the rejectable quality level in manufacturing RQL_m). Only then will the probability of lot failure in manufacturing be small. This can be demonstrated by making validation RQL equal to manufacturing AQL, namely, $RQL_v = AQL_m$. In this case, if validation passes, it will mean that the true fraction defective $p \leq RQL_v = AQL_m$. The validation AQL is set equal to or greater than CQL, namely, $AQL_v \geq CQL$. An estimate of CQL is obtained from prior characterization or other engineering studies as explained before.

This sequence of selections may be shown as follows:

Manufacturing		AQL_m	RQL_m
		$\Downarrow$	
Validation	AQL_v	RQL_v	
	$\Uparrow$		
Characterization	CQL		

Consider the example where $AQL_m = 1$ percent, $RQL_m = 5$ percent with the designed acceptance sampling plan being $n = 132$ and $c = 3$ as discussed before. This is the acceptance sampling plan to be used in manufacturing. If the prior estimate of CQL is 0.2 percent, then validation sampling plan will be designed with $AQL_v = 0.2$ percent and $RQL_v = 1$ percent. The designed sampling plan turns out to have $n = 666$ and $c = 3$. For an attribute sampling plan, if three lots are to be produced during validation, 222 items may be randomly selected from each lot and if the total number of defectives are 3 or less, validation passes.

It is not surprising that validation requires a larger sample size than lot acceptance testing. Validation accepts the process, lot acceptance tests accept a lot.

6.5 QUESTIONS TO ASK

To guide the proper design of acceptance sampling plans and process validation studies, here are some practical questions to ask:

1. Which type of sampling plan should be used, an attribute plan or a variable plan? The sample sizes for variable plans are smaller, but variable plans require actual measurements of the product and a normal distribution for the measured

characteristic. If the desired AQL and RQL are very small, there may not be any alternative except to use a variable plan.

2. If a sampling plan has been dictated by the customer or has otherwise been already proposed:
   - What is the operating characteristic curve of the plan?
   - Does the plan provide adequate protection to the consumer?
   - What information is available regarding the current quality level (CQL)?
   - In view of the CQL, does the plan provide adequate protection to the producer?

3. If an acceptance sampling plan is to be developed either for internal use or to provide to a supplier:
   - Has a risk assessment been done to determine criticality? What should be the AQL and RQL based upon the risk assessment?
   - What guidance is available as industry practice?
   - Are switching schemes going to be used as a part of the system of product acceptance?
   - What is the estimated CQL?
   - Is the estimated CQL less than the proposed AQL?
   - Is it practical to select the AQL based upon minimizing the total cost?

4. Is the possibility of using acceptance control charts being considered? These are control charts where periodic samples are taken to plot the charts, and the sample size and sampling interval are chosen such that when the production of the lot is complete, the number of samples on the control chart equals the number necessary for acceptance sampling. In this manner, SPC and acceptance sampling plans can be combined.

5. How are the validation sample sizes and acceptance criteria currently being established?

6. Are validation studies being designed with $AQL_v = CQL$ and $RQL_v = AQL_m$? If not, what is the justification for the approach being used?

Further Reading The following references provide further information on the topics discussed in this chapter. The ANSI standards provide tables of attribute and variable acceptance sampling plans along with their operating characteristic curves. They also include switching schemes. There are a large number of textbooks on acceptance sampling, such as Grant and Leavenworth (1980). The book by Joglekar (2003) provides methods to estimate CQL for both attribute and variable data.

CHAPTER 7

MANAGING AND IMPROVING PROCESSES: HOW TO USE AN AT-A-GLANCE-DISPLAY?

Statistical process control is widely used in industry. However, control charts are sometimes used without realizing that they are useful only if the process exhibits certain behavior. Control charts are often implemented without understanding or considering the risk and cost implications of the selected chart parameters. Also, the quarterly quality reviews are often inefficiently and ineffectively done. This chapter explains the fundamental rationale behind the development of control charts. It provides practical guidance to select subgroup size, control limits, and sampling interval. And it provides an at-a-glance-display of capability and performance indices to make it easy to plan, monitor, review, and manage process improvements.

This chapter assumes that you are familiar with statistical process control (SPC). If you are not, the subject is well covered in Joglekar (2003). This chapter is focused on SPC-related issues in industry where useful improvements can be made in implementing SPC. These topics include the practical and statistical considerations to select control limits, subgroup size, and sampling interval, and the use of capability and performance indices to plan quality improvements. The following four topics are considered:

1. *The Logic of Control Limits:* The lack of awareness of the logic behind the control limits can result in mistakes such as determining the $\bar{X}$ chart control limits based upon the standard deviation of all collected data, the plotting of specification limits on the $\bar{X}$ chart, or concluding that if the $\bar{X}$ chart control

Industrial Statistics: Practical Methods and Guidance for Improved Performance   By Anand M. Joglekar
Copyright © 2010 John Wiley & Sons, Inc.

limits are narrow the process variability must be small. The logic of control limits is explained.

2. *Selecting Subgroup Size:* Subgroup size is often arbitrarily selected. Software packages generally are not able to provide much assistance. The risk-based approach to select subgroup size is explained.

3. *Selecting Sampling Interval:* The selection of sampling interval is often arbitrarily done. Software packages generally are not able to provide much assistance. The practical considerations involved in selecting the sampling interval and a cost-based approach to select the sampling interval are described.

4. *At-A-Glance-Display of Capability and Performance Indices:* An at-a-glance-display of capability and performance indices is developed, which makes it easy to do quality planning, namely, to decide where control chart is the right tool to use, where other statistical tools such as DOE may be necessary, and where capital investments may have to be made. The display is a useful tool to monitor, review, and manage processes. It can significantly improve quarterly quality reviews.

7.1 STATISTICAL LOGIC OF CONTROL LIMITS

Let us suppose that you measure your body temperature by placing a thermometer under your tongue every hour. The temperature varies throughout the day due to a large number of factors. If the temperature variation is due to the large number of small common causes, then temperature is likely to have a normal distribution. The mean and standard deviation of temperature can be estimated from the collected data. Similarly, a process has common cause variation, which is likely to be normally distributed with a constant mean and standard deviation. If a process is only affected by common causes, it is said to be stable or in-control.

Now, suppose you have to find out if you have fever. Fever is defined as body temperature that is outside the range of the usual or the common cause variation. Fever is due to a special cause, and depending upon its magnitude, corrective actions may be necessary. Similarly, a process can exhibit special cause variation, and when it does so, it is said to be unstable or out-of-control. If this special cause variation is detected, it may be possible to find the cause and correct it. These actions help reduce the overall variability of the process and improve the quality of products produced.

Two mistakes can occur in making these judgments regarding whether the observed variation is due to common and special causes: common cause variation may be mistaken to be special cause and special cause variation may be mistaken to be common cause. If the type of variation is misdiagnosed, then wrong corrective actions may be taken. This is akin to taking aspirin when none is required and not taking it when required. These wrong corrective actions increase variability. The purpose of a control chart is to tell us when special causes are present so that the right corrective actions may be taken. Control limits provide the boundary to distinguish between common and special cause variations. If an observation falls outside the control limits, the presence of a special cause(s) is detected.

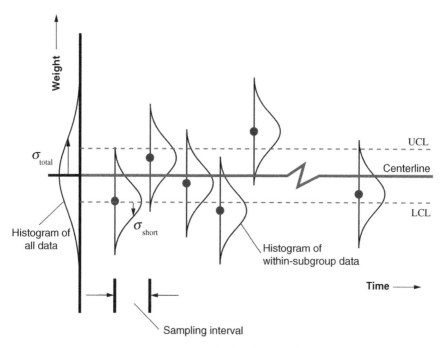

FIGURE 7.1   The logic of control limits.

What is the logic used to calculate control limits? Let us consider an example. Product weight is often a characteristic of interest. We have to find out whether our process is stable with respect to this characteristic. For this purpose, every hour, we measure the weights of n consecutive products produced by the manufacturing process. The data collected every sampling interval are known as a subgroup, the subgroup size being n. The sampling interval is 1 hour. For each subgroup, the subgroup average, standard deviation, and histogram may be generated and plotted over time as shown in Figure 7.1. The subgroup averages are shown by large dots. The within-subgroup standard deviation, also called the short-term standard deviation, is denoted by σ_{short}. This is the common cause variation of the process because the probability of a special cause affecting the process in a small duration of time over which the subgroup is collected is small. The standard deviation calculated from all collected data over a long time period is denoted by σ_{total}. This long-term standard deviation includes the effect of process instability caused by special causes. Usually, σ_{total} will be larger than σ_{short}. If the process is perfectly stable, the two standard deviations will be equal.

If the data suggest that both the mean and short-term standard deviation of the process are constant over time, then no special causes are present. Two control charts are created, the $\bar{X}$ chart to monitor process mean and the R (range) chart to monitor short-term process variability. For our purpose here, let us only consider the $\bar{X}$ chart. The same logic applies to all control charts.

A control chart has a centerline and control limits. Where should the centerline be? If the purpose of the control chart is to determine whether the process is stable, then the centerline should be drawn at the grand mean of all data denoted by $\bar{\bar{x}}$. On the other hand, if the purpose is to bring the mean on target, then the centerline should be at target. In Figure 7.1, the centerline is drawn at the grand mean.

How far away from the centerline should the control limits be? The function of the control limits is to help us identify when the process is out of control. The control limits should be drawn such that if the plotted $\bar{x}$ values fall randomly within the control limits, the process could be interpreted to be in control. If a plotted $\bar{x}$ value falls outside the control limits, the process will be deemed to be out of control at that point. The logic to compute the control limits is as follows:

1. The upper control limit (UCL) and the lower control limit (LCL) should be symmetrically positioned around the centerline. This is so because $\bar{x}$ is likely to have a symmetric distribution, usually the normal distribution.

2. The distance between each control limit and the centerline should be 3 sigma, because for a stable process the probability that a specific observed $\bar{x}$ will fall outside the control limits will then be less than 0.3 percent or 3 in 1000. If a plotted $\bar{x}$ falls outside the control limits, we will interpret this unusual event as an out-of-control signal rather than saying that 3 in a 1000 event has just occurred. Since 3 out of a 1000 plotted $\bar{x}$ values are likely to fall outside the control limits by pure chance for an in-control process, a false alarm will occur on the average once every 350 points or so. If this frequency of false alarm is deemed excessive, we can use (say) 4-sigma limits. The false alarm frequency will drop dramatically. However, the control limits will be wider and it will become more difficult to identify shifts in the process. Consequently, 3-sigma limits have now become a standard for all control charts.

3. What is sigma? Since we are plotting $\bar{x}$, sigma must be the standard deviation of $\bar{X}$. We know that $\sigma_{\bar{X}} = \sigma/\sqrt{n}$, where σ is the standard deviation of individual values and n is the subgroup size. So, the distance between UCL and the centerline equals $3\sigma/\sqrt{n}$.

4. How should sigma be computed? It needs to be computed to only reflect the common cause variability of the process. Only then will the control limits correctly separate common cause variation from special cause variation. σ_{short} denotes the short-term or common cause variability of the process (Figure 7.1). It would be wrong to compute it as the standard deviation of all data, denoted in Figure 7.1 by σ_{total}, because the process shown in Figure 7.1 may be unstable and σ_{total} could include both common and special cause variabilities. An estimate of σ_{short} can be computed from each subgroup and the pooled estimate obtained from all subgroups is taken to be the σ_{short}. More than 100 degrees of freedom are necessary to estimate any standard deviation including σ_{short} within 10 percent of the true value (see Joglekar (2003)). Thus, the control limits are as follows:

$$\text{Control limits for } \bar{X} \text{ chart} = \bar{\bar{x}} \pm 3\frac{\sigma_{\text{short}}}{\sqrt{n}} \qquad (7.1)$$

Note that four separate concepts have been used in determining control limits. The limits are symmetric around the mean because the distribution of $\bar{X}$ is symmetric. This is not necessarily the case with all other charts. For example, the control limits for the range chart are not symmetric, because the distribution of range is not symmetric. The multiplier 3 is used because we have decided to accept a false alarm rate of approximately 3 in 1000. σ_{short} is used because the limits must be based upon common cause variation. $\sqrt{n}$ is used for the $\bar{X}$ chart because the standard deviation of the average is smaller than that of individual values by a factor of $\sqrt{n}$.

As subgroup size n increases, control limits narrow inversely proportional to $\sqrt{n}$. The mere fact that the $\bar{X}$ chart control limits are inside the specification limits does not mean that the process is good. Specification limits are for individual values and $\bar{X}$ chart control limits are for averages. The control limits can be made narrow to any extent by simply increasing the subgroup size. It is usually wrong or counterproductive to draw control limits and specification limits on the same chart for both statistical and psychological reasons.

Commonly Used Control Charts   For variable data average and range, average and standard deviation and individual and moving range charts are used. For attribute data, p and c charts are used. Variable charts come in pairs and attribute charts come singly because for variable data, mean and standard deviation are independent of each other and need to be tracked separately. For attribute data, standard deviation is a deterministic function of the mean and need not be separately monitored. Special situations arise where modified limit chart, moving average chart, and short-run charts become useful. All of these charts and the key success factors necessary to implement the charts are described in detail in Joglekar (2003). A brief summary follows.

The average and range, or the average and standard deviation charts are used for variable data when the subgroup size is greater than one. The average and standard deviation charts are more efficient, but not by much, and the average and range charts continue to be more popular because of their familiarity and ease of computation. The individual and moving range charts are used for variable data with a subgroup size of one. The p chart is used when products are classified as defective or nondefective. The c chart is used when defects per product are counted.

Special situations arise. When the process capability is high, it may be costly to take corrective actions every time a point falls outside the standard control limits. Similarly, the larger the cost of corrective actions, the smaller the cost of scrap and rework, the wider the control limits should be from a cost perspective. This is the case of modified control limit charts described in Joglekar (2003) and the case of cost-based charts described later in this chapter. Another special case is when it is important to detect small changes in the mean of the process. The standard control charts are good at detecting large shifts in the mean. A moving average chart is better at detecting small shifts. Another common situation is when different products are produced using the same manufacturing process. Thus, product A is produced for some period of time, then the production shifts to product B, then to product C, then back to product B, and so on. In this case, instead of keeping a control chart for each product, it is possible to monitor the process using short-run charts.

The Case of the Smiling Faces   The daily production data (products produced per day), shown in Table 7.1, were displayed inside the entrance to a manufacturing plant. By the side of some of the numbers, smiling faces were drawn by the management. It turned out that the smiling faces were there when the daily production exceeded the goal of 3000 products per day. When the production workers were asked about the message conveyed by the smiling faces, their response was that management was trying to communicate that if everybody worked as hard every day as they had worked on the smiling face days, there will be smiling faces everywhere!

What can we learn from the data? We can construct a control chart from the collected data. The average daily production is $\bar{x} = 2860$. There is only one data point available per day or the subgroup size $n = 1$. The time series plot of daily production is called a chart of individuals. The short-term variance, namely, the within

TABLE 7.1   **Daily Production Data**

Day	Daily Production	mR
1	2899	
2	3028— ☺	129
3	2774	254
4	2969	195
5	3313— ☺	344
6	3139— ☺	174
7	2784	355
8	2544	240
9	2941	397
10	2762	179
11	2484	278
12	2328	156
13	2713	385
14	3215— ☺	502
15	2995	220
16	2854	141
17	3349— ☺	495
18	2845	504
19	3128— ☺	283
20	2557	571
21	2845	288
22	3028— ☺	183
23	2749	279
24	2660	89
25	2604	56

$\bar{x} = 2860$ and $m\bar{R} = 279$.

subgroup variance necessary to define control limits cannot be computed since the subgroup size is one. To assess short-term variability, the shortest term available for these data is one day. So a moving range mR (difference between successive observations) is calculated as shown, the average moving range being $m\bar{R} = 279$. The variability between consecutive days is considered to be the short-term variability and the short-term standard deviation (see Joglekar (2003)) and control limits are estimated as

$$\sigma_{\text{short}} = \frac{m\bar{R}}{1.128}$$

$$\text{Control limits for } X \text{ chart} = \bar{x} \pm 3\sigma_{\text{short}} = \bar{x} \pm 2.66 \, m\bar{R} \qquad (7.2)$$

For $\bar{x} = 2860$ and $m\bar{R} = 279$, we have centerline $= 2860$, LCL $= 2118$, and UCL $= 3602$. The X chart is shown in Figure 7.2. It shows the process to be stable. This means that the daily production could fluctuate between 2118 and 3602 products per day purely due to common cause variability. On some days, the goal of 3000 products is met. This happens about 30 percent (7 smiling faces out of 25) of the time. There are no special causes. There is nothing special about the smiling face days. It is not likely that the production workers worked harder on the smiling face days. Chances are that it is exactly the opposite. It is likely that on the smiling face days, when the daily production was large, everything worked well. Machines did not break down, good quality raw material was available, and so on. Perhaps people worked the least strenuously on those days. What is needed is an analysis of the entire process including suppliers to identify the causes of variability, and then, specific actions to improve the process. This is the responsibility of the management.

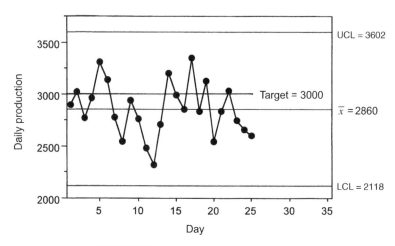

FIGURE 7.2 X chart for daily production.

7.2   SELECTING SUBGROUP SIZE

Let us now turn to the question of selecting a subgroup size. A key consideration in selecting a subgroup size is the risk of not detecting special causes. Risk-based determination of subgroup size for $\bar{X}$ chart follows. Similar considerations apply to all commonly used charts with subgroup sizes greater than one (Joglekar, 2003).

In making decisions with a control chart, there are two risks of wrong decision. These α and β risks are defined below:

α = Probability of concluding that a process is out of control when it is not
β = Probability of concluding that a process is in control when it is not

The α risk leads to false alarms and wasted efforts to detect the causes of process shift when no shift has occurred. With the usual design of control charts, the α risk is fixed at 3 in a 1000. The β risk implies inability to detect process shifts when they have occurred, causing larger off-target product to be produced. The β risk can be controlled by a proper selection of subgroup size.

The determination of subgroup size is now illustrated using a single point outside the control limits as the out-of-control rule. Figure 7.3(a) shows the α and β risks for the X chart (subgroup size $n = 1$) with 3σ limits. The α risk is $0.135 + 0.135$ percent = 0.27 percent. If the process mean shifts by an amount Δ, shown equal to 2σ as an illustration in Figure 7.3(a), then the process is clearly out of control. However, the probability that the plotted point immediately after the shift will fall

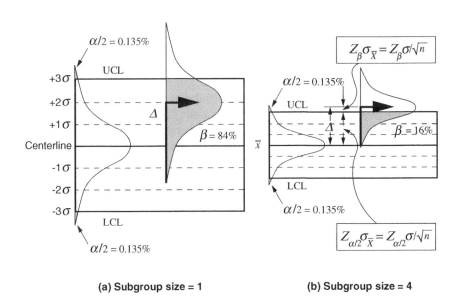

(a) Subgroup size = 1   (b) Subgroup size = 4

FIGURE 7.3   (a) and (b) Determining subgroup size.

inside the control limits is 84 percent. This is the β risk. In this case, there is an 84 percent chance of not immediately detecting a shift as big as 2σ. If the special cause produces a sustained shift in the mean of magnitude $\Delta = 2\sigma$, then the probability of not detecting it in k subsequent subgroups is $(0.84)^k$. Thus, for k equal to 2, 4, 6, 8, and 10, the β risk becomes progressively smaller, respectively, taking values of 70, 50, 35, 25, and 17 percent. At some point in time after the sustained shift has occurred, it is detected.

Figure 7.3(b) shows the effect of increasing the subgroup size. As subgroup size increases from one to four, the standard deviation of $\bar{X}$ reduces by a factor of 2 and the 3-sigma control limits tighten and become half as wide as they were before. The α risk remains unchanged because the control limits in Figure 7.3(b) are still 3-sigma limits. For the same shift Δ, the β risk reduces from 84 to 16 percent. Thus, for a fixed α risk, increasing the subgroup size reduces the β risk. By properly selecting the subgroup size, any desired β risk can be attained.

With reference to Figure 7.3(b),

$$\text{UCL} = \bar{\bar{x}} + Z_{\alpha/2}\frac{\sigma}{\sqrt{n}} = (\bar{\bar{x}}+\Delta) - Z_\beta \frac{\sigma}{\sqrt{n}}$$

Hence,

$$n = \left(\frac{Z_{\alpha/2}+Z_\beta}{d}\right)^2 \quad \text{where} \quad d = \frac{\Delta}{\sigma} \tag{7.3}$$

The σ in Equation (7.3) is σ_{short}. For the traditional 3-sigma limit charts, $Z_{\alpha/2} = 3$ and the subgroup size is as follows:

$$n = \left(\frac{3+Z_\beta}{d}\right)^2 \quad \text{where} \quad d = \frac{\Delta}{\sigma_{\text{short}}} \tag{7.4}$$

Table 7.2 shows the approximate values of subgroup size n for various values of d and β for 3-sigma limit $\bar{X}$ charts. For example, for $d=2$ and $\beta=25$ percent, the subgroup size $n=4$. $d=2$ means a shift of magnitude $\Delta = 2\sigma_{\text{short}}$, and β is the probability of not detecting a shift of magnitude Δ on the first subgroup after the shift.

TABLE 7.2 Subgroup Size n ($\alpha = 0.27$ Percent)

d	$\beta = 50\%$	$\beta = 25\%$	$\beta = 10\%$	$\beta = 5\%$
3.0	1	2	2	3
2.5	2	2	3	4
2.0	3	4	5	6
1.5	4	6	8	10
1.0	9	14	18	22

What about the ability of the $\bar{X}$ chart to detect sustained shifts? The probability that a shift will be detected on the kth subgroup following the shift is $\beta^{k-1}(1-\beta)$. Therefore, the expected number of subgroups to detect a sustained shift is as follows:

$$\sum_{k=1}^{\infty} k\beta^{k-1}(1-\beta) = \frac{1}{1-\beta} \qquad (7.5)$$

The necessary subgroup size can be determined if the magnitude of the sporadic shift to be detected immediately after it occurs, or the magnitude of the expected number of subgroups within which a certain sustained shift should be detected are specified.

One consequence of Equation (7.4) is that if σ_{short} is small, namely, the process capability is high, then for any fixed value of Δ the value of d will be large and the necessary subgroup size will be small. In addition to the statistical considerations described here, cost should also be formally or informally considered in selecting the subgroup size. For example, if the cost of sampling is high and the cost of scrap or salvage is low, subgroup size should be small.

7.3 SELECTING SAMPLING INTERVAL

Selecting the appropriate sampling interval is as important a decision as selecting control limits and subgroup size. Sampling interval influences the speed with which process shifts are detected and also affects the cost of sampling. We first discuss the practical considerations involved in selecting the sampling interval qualitatively, and then, describe a cost model approach to select the sampling interval.

1. An important factor is the expected frequency of process shifts requiring corrections. The smaller the interval between process shifts, the smaller the sampling interval should be.
2. If the process capability is large, that is, the chance of producing an out-of-specification product is small, then the process may be monitored infrequently. This is particularly so if the cost of sampling and testing is high.
3. If the cost of scrap and salvage is low, the sampling interval can be large.
4. The sampling interval can be event-driven and made to coincide with process changes likely to cause a process shift. Examples include shift changes, operator changes, raw material lot changes, and so on.
5. It may be desirable to sample frequently initially until a sufficiently large amount of data is collected and sufficient process knowledge is gained to obtain a better estimate of sampling interval.

Let us now consider a cost model proposed by Taguchi et al. (1989) and how it provides guidance with respect to the selection of control chart parameters. The model assumes a centered process. It is also assumed that a control chart with a certain subgroup size, sampling interval, and control limits is currently in place and that we want to optimize the sampling interval and control limits keeping the subgroup size the same as before. First, some nomenclature:

T = target for the product characteristic
Δ = half-width of specification
A = scrap or rework cost
B = measurement cost per subgroup
C = adjustment cost
N_0 = current sampling interval in number of units
N = sampling interval in number of units
D_0 = current control limit
D = control limit
l = time lag of measurement in number of units
u_0 = current average number of units between successive adjustments
$\bar{u}$ = predicted average number of units between successive adjustments
L = total quality cost per unit
σ_m^2 = measurement variance

Then, the various costs per unit produced are as follows:

$$\text{Measurement cost} = \frac{B}{N}$$

$$\text{Adjustment cost} = \frac{C}{\bar{u}}$$

As explained in Chapter 5, the loss per unit due to variability is given by

$$L = \frac{A}{\Delta^2}(y-T)^2$$

The average loss per product is given by $(A/\Delta^2)\sigma_y^2$, where σ_y^2 is the sum of the following three components:

- If the process is found to be in control, then assuming a *uniform* distribution, the $\sigma_y^2 = D^2/3$, the variance of the uniform distribution.
- If the process is found to be out of control, which will on average happen every $\bar{u}$ units produced, then $(N + 1)/2$ plus the lag l (in units) will be the average number of units approximately a distance D away from the target. This will cause an additional σ_y^2 per product equal to:

$$\left(\frac{N+1}{2}+l\right)\frac{D^2}{\bar{u}}$$

- Measurement error will cause an additional σ_y^2 equal to σ_m^2.

Adding all elements of cost, the total quality cost per unit is the following:

$$L = \frac{B}{N} + \frac{C}{\bar{u}} + \frac{A}{\Delta^2}\left[\frac{D^2}{3} + \frac{D^2}{\bar{u}}\left(\frac{N+1}{2}+l\right) + \sigma_m^2\right] \qquad (7.6)$$

The process is now assumed to behave in a manner that the average time taken for the characteristic to deviate a certain distance is proportional to distance squared. This implies that

$$\bar{u} = \frac{D^2}{D_0^2}u_0$$

Substituting $\bar{u}$ in Equation (7.6), differentiating Equation (7.6) with respect to N and D, and setting the derivatives equal to zero gives the optimum values of the sampling interval and control limits as follows:

$$N^* = \frac{\Delta}{D_0}\sqrt{\frac{2u_0 B}{A}} \qquad (7.7)$$

$$D^* = \left(\frac{3CD_0^2\Delta^2}{Au_0}\right)^{1/4} \qquad (7.8)$$

And the predicted P_p index (see the next section) is given by

$$P_p = \frac{2\Delta}{6\sqrt{\frac{D^{*2}}{3} + \left(\frac{n^*+1}{2}+l\right)\frac{D^{*2}}{\bar{u}} + \sigma_m^2}} \qquad (7.9)$$

Let us consider an example. Temperature is an in-process characteristic that has an effect on the dimension of the product produced. The specification for the dimension is $\pm 30\ \mu$m. A one-degree change in temperature produces a 5-μm change in the dimension. Hence, the specification for temperature is ± 6 degrees. The scrap cost is \$0.25. The cost of measuring the temperature is \$1.00 and the cost of making an adjustment to the process is \$2.00. The average sampling interval is approximately every 500 products produced and the average adjustment interval is approximately every 5000 products produced. The current control limits for temperature are ± 2 degrees. What should be the optimal sampling interval and control limits? What are the savings per product produced?

The above description translates to $\Delta = 6$, $A = 0.25$, $B = 1.00$, $C = 2.00$, $N_0 = 500$, $D_0 = 2$, $l = 0$, and $u_0 = 5000$ with the measurement standard deviation being essentially zero. Substituting these values in Equations (7.7) and (7.8) gives the optimum sampling interval to be 600 and the optimum control limits to be ± 0.9. From Equation (7.6), the quality control cost under the existing control scheme is \$0.013 per product. For the optimum control scheme, the average number of products between successive adjustments $\bar{u} = 1000$ and the quality control cost per product is \$0.007. Hence, the cost savings per product is \$0.006. From Equation (7.9), the P_p index will be 2.8.

Taguchi et al. (1989) provide cost models for both variable and attribute characteristics. They illustrate with examples how the various equations can be used to make resource allocation decisions on an economic basis. Even when all the assumptions are not applicable, or even if the approach is not formally adopted, it does provide the basis for determining the factors that influence the selection of sampling interval and control limits. For example, Equation (7.7) shows that wider specifications (higher performance index), larger measurement cost, and smaller scrap cost lead to longer sampling interval. From Equation (7.8), for wider specifications, higher correction cost, and smaller scrap cost, the economic action limits are wider. This substantiates the practical recommendations made earlier.

7.4 OUT-OF-CONTROL RULES

Out-of-control means that the process is not the same as it was. The change that has occurred may be good or bad. The purpose of the control chart is to detect the change. One rule to detect the change is when a single point falls outside the control limits. There are eight commonly used out-of-control rules to detect if the process is in control or not. The basic approach is that if a pattern of points has a very low probability of occurrence under the assumption of statistical control and can provide a practically meaningful interpretation of the likely nature of the special cause, then that pattern constitutes a test for special cause. Each of these eight tests has a probability of false alarm in the range of 3–5 in a 1000. Therefore, all the eight tests should not be simultaneously applied, otherwise these probabilities of false alarms will essentially add up to an unacceptable level of approximately 3 in a 100, that is, a false alarm may occur every 30 plotted points on average.

Test 1: Special cause is indicated when a single point falls outside the 3-sigma control limits. This suggests a sporadic shift or the beginning of a sustained shift.

Test 2: Special cause is indicated when at least two out of three successive points fall more than 2 sigma away on the same side of the centerline. This test provides an early warning of a shift.

Test 3: Special cause is indicated when at least four out of five successive points fall more than 1 sigma away on the same side of the centerline. This test provides an early warning of a sustained shift.

Test 4: Special cause is indicated when at least eight successive points fall on the same side of the centerline. This pattern suggests a sustained shift.

Test 5: Special cause is indicated when at least six consecutive points are steadily increasing or decreasing. The pattern suggests an upward or downward time trend.

Test 6: Special cause is indicated when there are at least 15 points in a row within ±1 sigma from the centerline on both sides of the centerline. This pattern indicates a large reduction in variability. It may be that the variability has in fact reduced, as may happen, if the batch of raw material is exceptionally homogeneous. This cause will be good to know because perhaps it could be perpetuated. However, there could be other reasons as well. This pattern could result if the gage is faulty, if the within-subgroup data come from a systematic sampling from two or more distributions, if process variability has dramatically reduced, or if wrong control limits are used.

Test 7: Special cause is indicated when there are eight or more points in a row on both sides of the centerline with none in the ± 1-sigma range. This pattern indicates that observations within a subgroup come from a single distribution, but each subgroup comes from one of two or more distributions.

Test 8: Special cause is indicated when there are at least 14 points in a row alternating up and down. This pattern indicates the presence of a systematic factor, such as two alternately used machines, suppliers, or operators.

Additionally, if the structure of the data suggests the need to investigate other meaningful patterns, the data should be so examined. For example, if the same raw material is supplied by two suppliers, the data may be segregated to see if there is a difference between suppliers. Similarly, the data may be examined for differences between shifts, operators, test equipment, and so on.

7.5 PROCESS CAPABILITY AND PERFORMANCE INDICES

The concept of process capability is different from the concept of process stability. A process is stable and predictable if the distribution of quality characteristic does not change over time. A stable process may not necessarily be a good process. The goodness of the process is measured by process capability, which is the ability of the process to produce products that meet specifications. A process is said to be capable if essentially all the products to be produced are predicted to be within specifications on the basis of objective evidence regarding the performance of the process.

For a stable process, meaningful predictions can be made regarding the probability of products being out of specification in the future. Depending upon this probability, a stable process may be classified as capable, potentially capable, or incapable. For an unstable process, the future distribution of the characteristic cannot be predicted with confidence and it is not possible to predict if the products produced in the future will be within specification or not. However, even for unstable processes, the following distinction may be made. Under the assumption that the degree of instability will continue as in the past, an unstable process may be capable. Additionally, if an unstable process is made stable by removing special causes and by centering as necessary, the resultant stable process will either turn out to be capable or incapable. Thus, an unstable process may be said to be unstable but capable, unstable but potentially capable, or unstable and incapable. Process characterization is very useful to identify improvement actions. Capability and performance indices allow us to classify processes in this manner.

TABLE 7.3   Capability and Performance Index Assumptions

Assumption	C_p	C_{pk}	P_p	P_{pk}
Two-sided specification	✓		✓	
Centered process	✓		✓	
Stable process	✓	✓		
Normal distribution	✓	✓	✓	✓

Two capability indices and two performance indices are defined below. Table 7.3 summarizes the varying assumptions made in the computation of these indices. The assumption of normality is not required to compute the indices. However, to predict the fraction defective corresponding to any given value of the index, an assumption regarding the distribution of the characteristic needs to be made. Typically, a normal distribution is assumed for this purpose.

C_p Index   The C_p index assumes a stable and centered process and a two-sided specification. Figure 7.4 shows this process with short-term standard deviation σ_{short}. The $\pm 3\sigma_{short}$ distance around the mean μ is defined as the process width. So, process width equals $6\sigma_{short}$. The C_p index is defined to be:

$$C_p = \frac{\text{Specification width}}{\text{Process width}} = \frac{\text{USL} - \text{LSL}}{6\sigma_{short}} \tag{7.10}$$

where USL is the upper specification limit and LSL is the lower specification limit.

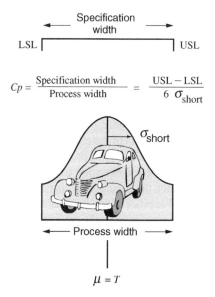

FIGURE 7.4   The car analogy.

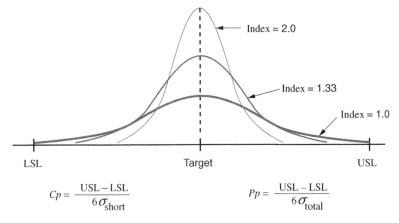

$$Cp = \frac{USL - LSL}{6\sigma_{short}} \qquad Pp = \frac{USL - LSL}{6\sigma_{total}}$$

FIGURE 7.5 C_p and P_p indices.

If the process width and the specification width are equal, as shown in Figure 7.4, then the C_p index is 1. If the process in Figure 7.4 was a fancy car and the specification was the garage, we will have a banged-up car with a C_p index of 1. A higher C_p index is desirable. What is the C_p index for a standard car on the roads in the United States? The C_p index in this case is the ratio of the width of the lane to the width of the car and is around 1.5. With this large C_p index, most of the time, most of us have little difficulty staying in our lane!

Figure 7.5 shows the distributions of individual values for C_p indices of 1.0, 1.33, and 2.0. A C_p index of 1 means that we have a single car that we wish to park in a one-car garage whose width is the same as the width of the car. A C_p index of 2 means that we have a single car that we wish to park in a two-car garage; a job that perhaps could be done with eyes closed. This is the target for 6-sigma quality. A process with a C_p index of 1.33 may be referred to as a 4-sigma process and may be acceptable. The larger the C_p index the better.

The C_p index can be improved in two ways: by widening the specification and by reducing the short-term variability.

C_{pk} Index From a practical viewpoint, the C_p index makes too many unrealistic assumptions. For example, if the specification is one-sided, the C_p index cannot be computed. Also, the process is assumed to be perfectly centered. This is rarely the case in practice. As shown in Table 7.3, C_{pk} index relaxes these two assumptions while retaining the assumption of stability.

Figure 7.6 shows a process where the mean is off-target. The C_{pk} index is defined as follows. If there is only one specification limit, only the relevant of the two computations below is done.

$$\frac{USL - Mean}{3\sigma_{short}} \quad \text{or} \quad \frac{Mean - LSL}{3\sigma_{short}}, \quad \text{whichever is smaller} \qquad (7.11)$$

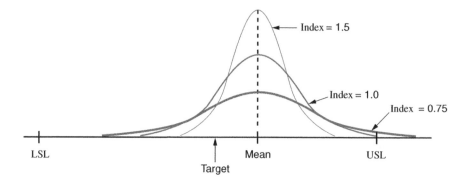

$$C_{pk} \text{ is the smaller of} \quad \begin{aligned} C_{pk} &= \frac{USL - Mean}{3\sigma_{short}} \\[2mm] C_{pk} &= \frac{Mean - LSL}{3\sigma_{short}} \end{aligned} \qquad P_{pk} \text{ is the smaller of} \quad \begin{aligned} P_{pk} &= \frac{USL - Mean}{3\sigma_{total}} \\[2mm] P_{pk} &= \frac{Mean - LSL}{3\sigma_{total}} \end{aligned}$$

FIGURE 7.6 C_{pk} and P_{pk} indices

The index may be improved in three ways: by widening the specification, by reducing the short-term variability, and by centering the process mean. When the mean is perfectly centered between a two-sided specification, C_{pk} index becomes equal to the C_p index.

P_p Index The P_p index makes the same assumptions as the C_p index, but with an important difference. It does not assume the process to be stable. As seen in Figure 7.5, and as defined below, σ_{total} is used to compute this index instead of σ_{short} The P_p index will be smaller than or equal to the C_p index because $\sigma_{total} \geq \sigma_{short}$.

The P_p index can be improved in three ways: by widening the specification, by controlling the process, which reduces special cause variability, and by reducing the short-term (common cause) variability.

$$P_p = \frac{USL - LSL}{6\sigma_{total}} \tag{7.12}$$

P_{pk} Index The P_{pk} index makes the same assumptions as the C_{pk} index with the important difference that the process is not assumed to be stable. As seen in Figure 7.6, and as defined below, σ_{total} is used to compute this index instead of σ_{short}. The P_{pk} index will be smaller than or equal to the C_{pk} index because $\sigma_{total} \geq \sigma_{short}$.

The P_{pk} index may be improved in four ways: by widening the specification, by centering the process mean, by controlling the process, which reduces the special cause variability, and by reducing the short-term (common cause) variability.

$$P_{pk} = \frac{USL - Mean}{3\sigma_{total}} \quad \text{or} \quad \frac{Mean - LSL}{3\sigma_{total}}, \quad \text{whichever is smaller} \tag{7.13}$$

The four indices are related as follows: $P_p C_{pk} = C_p P_{pk}$. Specifying any three indices determines the fourth. Also $P_{pk} < C_{pk} < C_p$ and $P_{pk} < P_p < C_p$. This understanding is used below to develop a very useful at-a-glace-display of these indices.

7.6   AT-A-GLANCE-DISPLAY

The four capability and performance indices permit an assessment of the process in terms of stability, centering, and capability. Figure 7.7 displays the four indices for a variety of situations. C_p, C_{pk}, and P_{pk} are shown by large dots. The three indices will always maintain the relationship $C_p \geq C_{pk} \geq P_{pk}$. The P_{pk} index measures the current performance of the process and the C_p index measures the best the process can do if centered and stabilized. The P_p index is shown by a dash, and it will be between P_{pk} and C_p indices. The key to read the display is shown in the right-hand portion of Figure 7.7. This figure can be used as an at-a-glance-display to plan, monitor, review, and manage quality improvements.

The values of the four indices can be easily computed from the data usually collected for variable control charts. This is done by estimating $\bar{x}$, s_{short}, and s_{total} from the collected data. Joglekar (2003) discusses these computations for a variety of

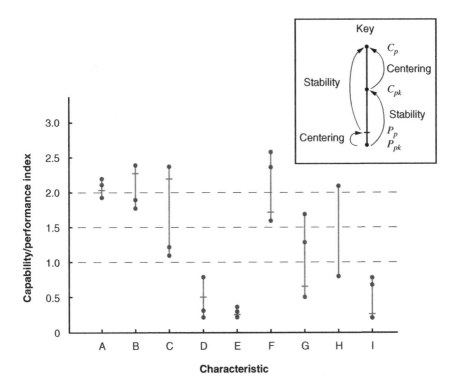

FIGURE 7.7   At-a-glance-display of capability and performance indices.

situations. Based upon the values of the process capability and performance indices, processes may be classified into the following six categories. The six categories are obtained by classifying the process to be stable or unstable, and then, as capable, potentially capable, or incapable.

Stable and Capable   These are stable processes whose current performance as measured by the P_{pk} index meets the target. The signature of this is shown in Figure 7.7, characteristic A, which shows a stable, centered, and capable process. If the process is not well centered, the signature may be as in Figure 7.7, characteristic B. In this case, the current performance meets the target, but could be improved further by centering the process. Stable and capable processes do not require significant improvement effort. These processes should be monitored and centered if necessary.

Stable and Potentially Capable   These are stable processes whose current performance as measured by the P_{pk} index does not meet the target, but will become satisfactory if the process is better centered. In other words while P_{pk} does not meet the target, P_p does. The signature of this process is shown in Figure 7.7, characteristic C. These processes need to be better centered and the process improvement plan should state how that will be done. Control chart is a useful tool in this case.

Stable and Incapable   These are stable processes that, even if centered properly, will not produce acceptable performance. For these processes, the C_p index is too low to be acceptable. The signature of these processes is shown in Figure 7.7, characteristics D and E. To expect a control chart to significantly improve these processes is unrealistic. If the only proposed strategy to improve these processes is to put them on a control chart discipline, the strategy is in error. The process needs to be changed to reduce short-term variability. This may require the use of tools such as variance component analysis, measurement system analysis, and DOE. If process improvements do not work, a different process and capital investments may be necessary. The process improvement plan needs to state the specific strategies to be used.

Unstable but Capable   These are unstable processes that are currently capable. The P_{pk} index meets the desired target. These processes may or may not be centered correctly. Figure 7.7, characteristic F shows one signature of this process. Even though the current levels of instability and noncentering are satisfactory, this process will benefit from being control charted, stabilized, and centered. If the process is made stable, it will provide greater assurance of continued good performance in the future.

Unstable but Potentially Capable   These are unstable processes whose current performance is unsatisfactory, but can be made satisfactory by improving stability and centering. Figure 7.7, characteristics G and H shows signatures of these processes. For Figure 7.7, characteristic G, $P_{pk} = 0.5$, $P_p = 0.65$, $C_{pk} = 1.3$, and $C_p = 1.7$. This means that the current performance of the process is unsatisfactory, but if the process could be stabilized and centered, it will become a capable process. For Figure 7.7, characteristic H, there are only two dots, that is, the specification is one-sided and only

P_{pk} (lower dot) and C_{pk} (upper dot) can be calculated. Again, the current performance is not satisfactory, but if the process is made stable it will become very capable. A control chart is extremely useful for these processes. What is necessary is to monitor the process using a control chart, find and remove special causes as they arise, and adjust the mean as necessary.

Unstable and Incapable These are unstable processes whose current performance is unsatisfactory and while the performance may be improved by stabilizing and centering the process, the improved process will still not be satisfactory. Figure 7.7, characteristic I is a signature of this process. The process is unstable, since P_{pk} and C_{pk} differ. The process happens to be well centered since C_p and C_{pk} are close together. However, even if the process is made stable, it would continue to be incapable since C_p is not satisfactory. By controlling and centering these processes, some improvement in performance is possible. Again, control charts are useful for this purpose. However, common cause variability will have to be reduced as well. Key causes of variability will have to be identified using tools such as variance component analysis, measurement system analysis, design of experiments, and so on. Or, a completely new process will have to be implemented.

Use of this at-a-glance-display is highly recommended for quality planning. Each product typically has multiple characteristics of interest and they could all be displayed on a single figure (Figure 7.7), with an attachment that identifies the improvement strategies for that product. This display becomes a communication tool to be used in quality reviews instead of a series of control charts as is often done. The display could be updated every quarter based upon new data to monitor, manage, and communicate improvements.

7.7 QUESTIONS TO ASK

The following are some questions to ask to improve the implementation of SPC:

1. For any particular product, what does the at-a-glance-display look like for all characteristics of interest?
2. Is the display based upon adequate amount of data (see Joglekar (2003))?
3. Based upon this display, what is the quality improvement plan for the product?
   - Which characteristics should be control charted to improve centering or stability?
   - Which characteristics need common cause variance reduction efforts?
   - Where is a complete process change necessary?
4. For characteristics to be control charted, is the correct chart being used? Are the control limits, subgroup size, and sampling interval determined using the guidance provided here?
5. If $\bar{X}$ chart is being used, what is the β risk associated with the selected subgroup size?

6. What cost considerations were used in selecting the control chart parameters?
7. Have the out-of-control action procedures been defined and communicated to the production operators?
8. Is the at-a-glance-display being used to make presentations at the quarterly quality reviews?
9. Does the display show that improvements are occurring quarter by quarter as planned? If not, what actions are necessary?

Further Reading   The book by Joglekar (2003) covers the subject of statistical process control in considerable detail including the commonly used and special control charts and an in-depth explanation of process capability and performance indices. It includes many practical applications. The book by Taguchi et al. (1989) covers the cost-based approach to process control and provides many practical applications of the approach.

CHAPTER 8

HOW TO FIND CAUSES OF VARIATION BY JUST LOOKING SYSTEMATICALLY?

Reducing variability is an important objective in manufacturing. Designed experiments, intended to identify causes of variability and so useful in R&D, are often not practical on a manufacturing scale. On the other hand, a large amount of passively collected data is available in manufacturing. Variance components analysis is an important tool to analyze such data to identify the key causes of variation and the contribution of each cause to the total variance. This chapter explains the basic principles of variance components analysis, how such an analysis can be done with data routinely collected in manufacturing, and how the results can be used to develop cost-effective improvement strategies.

Excessive product variability is detrimental to the health of a company. It degrades product quality and customer satisfaction. It can cause lots to fail in manufacturing leading to large financial losses. Consequently, reducing product variability is a key objective in industry. A large number of people are devoted to this task. Perhaps, this is an element of your job description as well.

Industry Examples A manufacturer of drug-coated tablets found that for a new product recently put into manufacturing, batches of tablets were occasionally produced where the *in vitro* release rate variability was much larger. Anecdotally, the scientists and engineers believed the cause to be the release rate measurement system. While the measurement system had been validated, it was felt that not all analysts were equally well trained in taking measurements and perhaps the measurement system was

Industrial Statistics: Practical Methods and Guidance for Improved Performance By Anand M. Joglekar
Copyright © 2010 John Wiley & Sons, Inc.

not very well controlled. People in the analytical department disputed this and were of the opinion that the problem lay squarely in the court of manufacturing. They needed to control the manufacturing process better. Considerable data had been routinely collected over the past two months since manufacturing began, including data on raw materials, in-process controls, final product characteristics, ambient conditions, and so on. The data had been stored away and not analyzed. Analysis of this passively collected data indicated ambient humidity to be the problem, and this was confirmed later by small-scale designed experiments.

A food company was experiencing large product-to-product variability. The manufacturing machines were old. Engineers in this food company believed that this old, in-house equipment was responsible for product variability. A capital investment proposal had been developed and was awaiting the signature of the chief financial officer. Meanwhile, some of the engineers involved in this project were in my seminar on variance components analysis. One requirement of this in-house seminar was that each team had to make an application of the tools taught in the seminar while the instructor was on-site and available for consultation. This team of engineers decided to reanalyze the available manufacturing data as a class project. The variance decomposition showed that the manufacturing machines were not responsible for the observed variation and over half-million dollars in capital expense was avoided.

A company produces a certain kind of paste and the typical manufacturing operation involves procurement of raw materials, manufacture of the product, and product testing prior to release. The moisture content of this paste is an important functional characteristic. The target value for moisture is 10 percent and the specification is from 9 to 11 percent. Figure 8.1 shows the histogram of 60 available moisture observations. The average is 10.27 percent, slightly above target. Whether the true population mean is on target may be assessed by constructing a confidence interval for the mean. Approximately 3.3 percent of the data is below the lower specification limit and 10 percent of the data is above the upper specification limit. With so much of the product outside the specification, the current situation needs to be improved.

Cost-effective variance reduction becomes easier if the key causes of variation are known. Otherwise, variance reduction efforts can easily be misdirected. These key causes of product variation may be found by using designed experiments where "active" or purposeful changes are made to factors likely to cause variation.

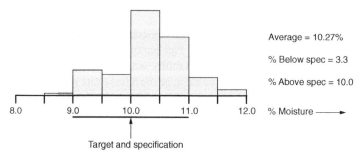

FIGURE 8.1   Histogram of moisture data.

Such purposeful changes are easier to make in the research and development phase, but difficult to make in manufacturing due to the large cost of conducting full-scale manufacturing experiments.

A second approach is to merely observe what is going on. No changes are being purposefully made, but "passive" changes are occurring on their own accord. This is almost always the case during manufacturing where a large number of factors (raw material lots, process settings, and many other factors) are changing and these changes are being recorded. Murphy is at work. The product characteristics of interest are changing as well and are also being recorded. How can we analyze such passively collected data to identify the key sources of variation? Is there a way to listen to Murphy speak?

8.1   MANUFACTURING APPLICATION

The Process   Let us consider an example from Joglekar (2003) based upon data reported in Box, Hunter, and Hunter (1978). The paste manufacturing process is shown in Figure 8.2. The process consists of feeding incoming batches of raw material into a continuous manufacturing process that ultimately produces drums of paste. Each batch of raw material, which is homogeneous within itself, produces 100 drums of paste. To monitor the process, 2 drums are randomly selected from the 100 drums produced from each batch of raw material, and two moisture measurements are made on samples taken from each drum. When this variance reduction project started, data had been collected on 15 raw material batches, for a total of 60 ($15 \times 2 \times 2$) observations.

What is meant by looking systematically is the recognition that each moisture observation is not just an isolated observation, but has an identification tag associated with it indicating the raw material batch and the drum of paste the observation came from. Such data are routinely collected in manufacturing. Most of the time, the collected data are merely filed away. To be more useful for analysis purposes, we need to systematically look for the associated tag.

Collected Data   The 60 moisture observations and the associated batches, drums, and tests are shown in Table 8.1. Figure 8.1 was a histogram of these moisture

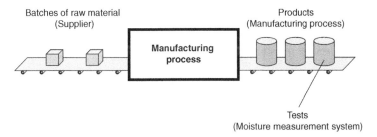

FIGURE 8.2   The manufacturing process.

TABLE 8.1   Moisture Data

Batch	Drum	Test	Moisture (%)	Batch	Drum	Test	Moisture (%)
1	1	1	11.6	8	2	1	10.5
1	1	2	11.1	8	2	2	10.8
1	2	1	10.6	9	1	1	10.3
1	2	2	10.8	9	1	2	10.4
2	1	1	10.2	9	2	1	10.7
2	1	2	10.4	9	2	2	10.7
2	2	1	10.1	10	1	1	8.9
2	2	2	10.2	10	1	2	9.2
3	1	1	10.5	10	2	1	10.3
3	1	2	10.3	10	2	2	10.0
3	2	1	9.0	11	1	1	10.1
3	2	2	9.2	11	1	2	9.9
4	1	1	11.2	11	2	1	10.0
4	1	2	11.4	11	2	2	10.3
4	2	1	10.3	12	1	1	10.5
4	2	2	10.6	12	1	2	10.6
5	1	1	9.5	12	2	1	10.7
5	1	2	9.7	12	2	2	10.8
5	2	1	9.3	13	1	1	9.2
5	2	2	9.1	13	1	2	9.3
6	1	1	10.9	13	2	1	10.2
6	1	2	10.8	13	2	2	10.1
6	2	1	10.2	14	1	1	9.9
6	2	2	10.0	14	1	2	10.0
7	1	1	9.5	14	2	1	10.1
7	1	2	9.6	14	2	2	10.4
7	2	1	10.5	15	1	1	11.5
7	2	2	10.4	15	1	2	11.3
8	1	1	11.0	15	2	1	10.2
8	1	2	10.8	15	2	2	10.4

observations. Note that drums 1 and 2 are not the first and second drums of paste produced from a specific batch of raw material; rather they are two randomly selected drums produced from the specific batch of raw material. Similarly, the two tests are labeled 1 and 2 only for convenience and are not the first and the second tests.

Objectives of Data Analysis   To cost effectively reduce variance, we need to answer the following questions.

1. Our first question has to do with deciding whether the current situation is satisfactory. There are many ways to do this. A quick answer is provided by the histogram in Figure 8.1 that helps answer the following questions: Is the mean moisture on target? How big is the moisture variability? What fraction of the product is out of specification? If the entire histogram is well within

specifications, we may not feel the urgency to make improvements. We could have computed the performance index or used tolerance interval approach to predict the expected out-of-specification product. This seems unnecessary at the moment because the situation is clear by just looking at the histogram.

2. If improvements are necessary, our second question has to do with identifying the causes of variability so that we can focus our improvement efforts correctly. Presently, there are three potential causes of variability, and each cause is referred to as a variance component. The three variance components are the following: (1) Moisture variability caused by raw material batch-to-batch variation. To reduce this component of variance, we will have to work with the supplier. (2) Moisture variability caused by drum-to-drum variation (product to product). To reduce this variance component, we will have to improve our manufacturing process. (3) Moisture variability caused by test-to-test differences. To reduce this variance component, we will have to improve the measurement process. If you are managing this project, you want to know whether you should call the supplier, the head of manufacturing, or the head of analytical department. What we specifically want to know is the percentage of the final moisture variance due to each of these three causes. Clearly, if internal manufacturing causes most of the variation in the moisture content of the product, then there is not much point in attempting to reduce the measurement system variability.

3. The final question has to do with setting improvement targets and allocating resources, namely, developing an action plan for improvement. If manufacturing variability is the largest source of variation, how much reduction in manufacturing variability is necessary to achieve the desired quality target? From a return on investment viewpoint, how much money can be spent to accomplish this objective? The answers to these questions help establish a practical improvement plan.

Data Structure   The data and the associated tags may be graphically represented as a tree diagram shown in Figure 8.3. The collected moisture data are in the bottom row of Figure 8.3. There were two moisture measurements taken on each drum of paste. The average of these two tests on the same drum represents our estimate of the mean

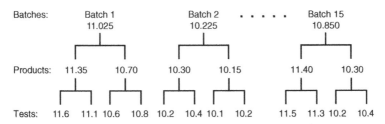

FIGURE 8.3   Data structure.

moisture in that drum. These drum moisture averages are shown in the middle row of Figure 8.3 labeled Products. Similarly, there were two drums per batch of raw material and the average of the corresponding four tests represents our estimate of the average moisture in all the drums produced from the same raw material batch. These batch moisture averages are shown in the top row of Figure 8.3 labeled Batches.

8.2   VARIANCE COMPONENTS ANALYSIS

From the collected data, we wish to estimate the following three variance components:

σ_B^2 = Moisture variability due to batches of raw material(supplier)
σ_P^2 = Moisture variability due to drum-to-drum(product-to-product)
    differences within a batch of raw material(manufacturing process)
σ_e^2 = Moisture variability due to test-to-test differences within a drum
    (measurement method)

The concept of variance components is graphically illustrated in Figure 8.4. The overall true mean moisture is represented by process mean. Let batch mean be the true average moisture in all the drums produced by a specific batch of raw material. Let product mean be the true moisture in a specific drum produced from this raw material batch. The observation is the test result on a sample from this particular drum. The three variance components add to produce the total moisture variability:

$$\sigma_t^2 = \sigma_B^2 + \sigma_P^2 + \sigma_e^2$$

where σ_t^2 is the total variance of moisture.

The three variance components may be estimated as follows:

1. *Test Variance:* With reference to Figure 8.3, for drum 1 made from the first batch of raw material, the two test results are 11.6 and 11.1. These results differ from each other only due to test variability. An estimate of test variance, from this one pair of results, is 0.125. This estimate has one degree of freedom. There

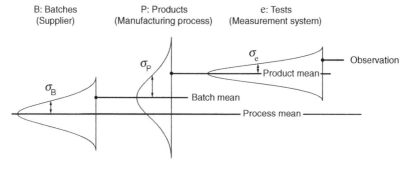

FIGURE 8.4   The concept of variance components.

are 30 such pairs of test results. The average of the 30 test variance estimates gives

$$s_e^2 = 0.023 \text{ as the estimated test variance with 30 degrees of freedom}$$

Note that because the moisture measurement test is a destructive test, this estimate of measurement variability necessarily includes the sample-to-sample variability within a drum.

2. *Product Variance:* For the first batch, there are two drum average moistures, 11.35 and 10.70, obtained by averaging the corresponding two test results. The variance calculated from these two averages is 0.06125. These two drum averages differ from each other not only due to drum-to-drum differences, but also due to test variability. Since two test results are averaged, 0.06125 estimates $\sigma_P^2 + \left(\sigma_e^2/2\right)$. There are 15 such pairs of averages and the pooled estimate of the variance of drum moisture averages is 0.2653. Hence,

$$s_P^2 + \frac{s_e^2}{2} = 0.2653$$

Substituting for s_e^2, we get

$$s_P^2 = 0.254 \text{ as the estimated product-to-product}$$
$$\text{variance with 15 degrees of freedom}$$

3. *Batch Variance:* There are 15 batch averages. The variance calculated from these batch averages is 0.266. These batch averages differ from each other because of batch-to-batch, product-to-product, and test-to-test differences. Since each batch average consists of two products and four tests, the variance calculated from the batch averages estimates $\sigma_B^2 + \sigma_P^2/2 + \sigma_e^2/4$. Hence,

$$s_B^2 + \frac{s_P^2}{2} + \frac{s_e^2}{4} = 0.266$$

and substituting for s_e^2 and s_P^2, we get $s_B^2 = 0.133$ as the batch-to-batch variance with 14 degrees of freedom.

This variance components analysis is summarized in Table 8.2. There are three sources (components) of variance. Column (a) shows the degrees of freedom for each variance component. Larger degrees of freedom provide a more precise estimate of the variance component. Column (b) shows the computed variance components due to each source and these component variances add to the total variance. Column (c) shows the standard deviation due to each source of variability and is the square root of the corresponding variance. Note that while the variances add to the total variance, standard deviations do not arithmetically add to the total standard deviation. Column (d) shows the percent contribution of each source to the total moisture variance obtained by expressing each component variance as a percentage of the total variance. Approximately, the supplier is responsible for 32 percent of moisture variability, the

TABLE 8.2 Variance Components

Source	(a) Degrees of Freedom	(b) Variance	(c) Standard Deviation	(d) Percent Contribution (%)
Batches (supplier)	14	0.133	0.365	32.4
Products (manufacturing process)	15	0.254	0.504	62.0
Test (measurement system)	30	0.023	0.152	5.6
Total	59	0.410	0.640	100.0

manufacturing process is responsible for 62 percent of moisture variability, and the measurement method accounts for 6 percent of the variance. This is precisely the information we need to plan cost-effective improvements.

8.3 PLANNING FOR QUALITY IMPROVEMENT

We now combine the tools of variance components analysis, process capability, and economic loss to develop cost-effective variance reduction strategies by conducting a series of "what if" analyses as shown in Table 8.3.

1. *Current Process:* Column (a) in Table 8.3 shows the results for the process as it currently is. The target moisture is 10 percent with a two-sided specification of 9–11 percent. The current process mean is 10.27 percent. There are three sources of variance: supplier variance is 0.133, manufacturing variance is

TABLE 8.3 Planning for Quality Improvement

	(a) Current Process	(b) Mean on Target	(c) Reduction in Manufacturing Variability (c1) 50%	(c2) 100%	(d) 90% Reduction in Supplier and Manufacturing Variability
Process mean	10.27	10.00	10.00	10.00	10.00
Supplier variance	0.133	0.133	0.133	0.133	0.013
Manufacturing variance	0.254	0.245	0.122	0.000	0.025
Test variance	0.023	0.023	0.023	0.023	0.023
Total variance	0.410	0.410	0.278	0.156	0.061
% out of specification	15.45	11.88	5.70	1.14	0.00
P_{pk} index	0.38	0.52	0.63	0.84	1.35
Classical loss per year ($k)	309	238	114	23	0
Quadratic loss per year ($k)	965	820	556	312	122

0.254, and test variance is 0.023. The total variance is 0.410, which gives $\sigma_t = 0.64$. Assuming normality, out-of-specification product is calculated to be 15.45 percent. The P_{pk} index is calculated as $(11.0 - 10.27)/(3 \times 0.64) = 0.38$.

The economic loss due to out-of-specification product may be calculated as follows based upon the classical scrap and salvage loss function. The cost of producing and finding an out-of-specification product is $20 per drum. The yearly sales are 100 thousand drums. Since 15.45 percent of the production is out of specification,

Classical loss per year $= 0.1545 \times 20 \times 100,000 = \$309,000.$

The economic loss based upon the quadratic loss function may be calculated once a single point on the loss function is specified. The economic loss to the customer at either specification limit is in fact greater than $20, but for purposes of direct comparison, let us assume the loss to be $20. This means that

Quadratic loss at specification limit $= k(11-10)^2 = 20$

Hence, $k = 20$.

Quadratic loss per product $= k[(\mu - T)^2 + \sigma^2] = 20(0.27^2 + 0.64^2) = 9.65$
Quadratic loss per year $= \$965,000$

The quadratic loss is considerably higher than the classical loss because the quadratic loss function assumes that economic losses occur even inside the specification.

The percent out-of-specification product, the P_{pk} index, and the projected yearly economic loss suggest the need for process improvement. The yearly economic loss also provides a feel for the resources that could be committed to improve this process. A formal return on investment analysis can be done. Had the projected economic loss turned out to be small, attention would have been focused on some other process where larger gains could be made.

2. *Mean on Target:* What if the process mean could be centered exactly on target? The results would be as shown in column (b) of Table 8.3. The percent out-of-specification product will reduce to 11.88; the P_{pk} index will improve to 0.52 and will now be the same as the P_p index. The classical economic loss will reduce to $238k. The quadratic loss will reduce to $820k.

3. *Reduced Manufacturing Variability:* Further improvements can only be achieved by reducing variability. The initial focus needs to be on reducing manufacturing variability since it is the largest variance component. What if the manufacturing variability could be reduced by 50 percent if certain investments are made?

Column (c1) in Table 8.3 shows the predicted results. The classical economic loss will reduce to $114k. The quadratic loss will be $556k. Along with centering, this improvement action is worth $200k to $400k depending upon the selected loss function. If an investment of $200k was deemed necessary to accomplish this objective, the investment will be recouped in less than one year.

Note that even with this projected improvement, the P_{pk} index is still only 0.63. The corporate goal may be a P_{pk} of 1.33. Can we achieve a P_{pk} index of 1.33 by focusing on manufacturing process improvements alone? Column (c2) in Table 8.3 shows that even if the entire manufacturing process variability is eliminated, the P_{pk} index will only improve to 0.84.

4. *Reduced Supplier and Manufacturing Variability:* To achieve a P_{pk} index of 1.33, a 90 percent reduction in both the supplier and manufacturing variance is necessary as shown in column (d) of Table 8.3. The out-of-specification product and the classical economic loss essentially become zero. The quadratic loss becomes $122k, which is a reduction of over $800k. Expenditures in the range of $300k to $800k are easily justifiable, depending upon the selected loss function, to accomplish all of these improvements.

8.4 STRUCTURED STUDIES

Variance components analysis is typically conducted using a software package. However, the software package needs to know the structure of the study: whether the factors in the study are fixed or random factors, and which factors are nested and which factors are crossed. Most of the software packages of today assume all factors to be random factors and this necessitates a correction to be applied to the computed variance components for fixed factors.

Fixed and Random Factors Whether a factor is a fixed factor or a random factor depends upon the way the levels of the factor are selected and the inferences to be drawn from the analysis. A factor is fixed if the inferences are limited to the chosen levels of the factor. In this sense, the entire population of levels is included in the experiment. As an example of a fixed factor, suppose a manufacturing plant has three similar machines to produce a product. If we are interested in determining the variability caused by differences between these three specific machines, then machine is a fixed factor. If there are five formulations of particular interest that an experimenter wishes to evaluate and these represent the entire set of formulations regarding which the experimenter wishes to make inferences, then formulation is a fixed factor.

A factor is random if the selected levels of the factor represent random drawings from a much larger (infinite) population. Inferences are to be made regarding the infinite population, not just the levels represented in the experiment. Raw material lots are an example of random factor. A few lots of raw material may be included in the study, but the interest centers not on the variability caused by these specific lots, but on the variability caused by the entire population of lots to be used in future production.

Sometimes it may be difficult to decide whether a given factor is fixed or random. The main difference lies in whether the levels of the factor can be considered to be random samples from a large population or represent the entire population of interest.

Consider a factor such as ambient humidity. Let us assume that this environmental factor has a large influence on the quality of the product produced. To investigate this effect, two widely different levels of humidity (say 20 and 70 percent) are chosen and appropriate experiments are conducted in humidity chambers. Is humidity a random factor or a fixed factor? The answer is that during production, humidity will vary throughout the year according to some distribution. So, during actual production, humidity is a random factor and we are ultimately interested in predicting the effect of this humidity distribution on product quality. However, in the experiment itself, two levels of humidity were specifically chosen to represent a wide range of humidity. Therefore, in the experiment, humidity is a fixed factor. Consider another example where the same factor may be a fixed factor for one company and random factor for another. Consider the case of machines used to analyze human blood chemistry. For the manufacturer of these machines, machine is a random factor because the manufacturer is interested in the variability caused by the entire population of machines sold. For a hospital that uses two similar machines to conduct the analysis, machines is a fixed factor because the interest of the hospital is in the variability caused by these two machines, and not in the variability caused by the other thousands of machines in the market.

As mentioned earlier, many software packages currently treat all factors as random factors. The effect of treating a fixed factor as a random factor is to overestimate its variance contribution. If there is a fixed factor with k levels, then the variance contribution calculated assuming it to be random should be changed as follows (Joglekar, 2003):

$$\sigma^2(\text{fixed}) = \frac{k-1}{k}\sigma^2(\text{random})$$

Nested and Crossed Factors   Figure 8.5 shows a nested structure involving lots and products within a lot. Products are said to be nested inside lots. There are a lots. From each lot, b products are randomly selected for evaluation. Even though the products of each lot are labeled $1, 2, \ldots, b$, product 1 of lot 1 is different from product 1 of lot 2. In this sense, the b levels of product are different for each lot. If all product 1 results are averaged, this average does not have a practically different meaning compared to the average of, say, products labeled 2.

A nested structure or classification is a hierarchical structure and looks like a tree diagram. Levels of the nested factor are all different for each level of the prior factor. A nested structure may have multiple levels of nesting so that factor B may be nested

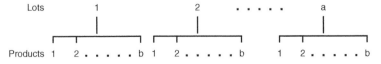

FIGURE 8.5   Nested structure.

Lanes	1	2	Times	b
1	X	X		X
2	X	X		X
. . . .				
a	X	X		X

FIGURE 8.6   Crossed structure.

inside A, C may be nested inside B, and so on. The factors themselves may be random or fixed. Lots and products are typically random factors. Instead, there may be a machines and b products per machine. Here, machines may be a fixed factor and products are random. If there were three similar machines with four stations per machine, both machines and stations would be considered as fixed factors. A nested structure allows the total variability to be partitioned into variability due to each factor.

Figure 8.6 shows a two-factor crossed structure. The two factors are lanes and times. There are multiple lanes of products being simultaneously produced as with multihead filling machines. There are a lanes and b times. What distinguishes a crossed structure from a nested structure is the fact that time 1 for lane 1 is exactly the same as time 1 for lane 2. Similarly, lane 1 for time 1 is exactly the same as lane 1 for all levels of time. In a crossed structure, data cells are formed by combining each level of one factor with each level of every other factor. The resultant structure looks like a matrix. In this case, the average result for time 1 has a practically meaningful and different interpretation from the average result for time 2.

A crossed structure may have more than two factors. The factors may be fixed or random. In the above example, lanes is a fixed factor and time is random. A crossed structure allows the total variability to be partitioned into variability due to each factor and variability due to interactions between factors. In the above example, lanes and times are said to interact if the difference between lanes changes with time. For a nested structure, if B is nested inside A, then A and B cannot interact. Chapter 9 on measurement systems analysis considers an example involving crossed and nested structures including the interpretation of the variance component due to interaction. When some factors are crossed and others nested, the structure is called a mixed structure. See Joglekar (2003) for many other applications of crossed and mixed structures.

Let us consider some additional examples. There are three operators who are charged with taking measurements on a production line. In a measurement system study, each operator took duplicate measurements on the same 10 randomly selected products from the production line. What is the structure of this study? Operator is a fixed factor. If the three operators were randomly selected from a large pool of operators, then operators would have been a random factor. Products and duplicates are random factors. This study has a mixed structure. Products and operators are crossed and duplicates are nested inside products and operators. Consider another example. A hamburger chain has five fast food restaurants. Five groups of 10 different people were selected and each group went to a different restaurant. Each person evaluated the quality of two hamburgers. What is the structure of this study? Restaurant is a fixed factor because the entire population of restaurants of interest is in the study. People and hamburgers are random factors. This is a nested study. A different group of people went to each restaurant; hence, people are nested inside restaurants. Since each person evaluated different hamburgers, hamburgers are nested inside people.

Selecting Degrees of Freedom   A key question in designing a variance components study is to decide the degrees of freedom for each variance component. For example, the three sources of variability may be lot-to-lot variability, product-to-product variability within a lot, and duplicate variability. How many degrees of freedom should be allocated to each of these three sources of variability? If the lot-to-lot variance component is to be estimated with four degrees of freedom, then there should be five lots in the study. Once the degrees of freedom are assigned to each variance component, the study can be designed.

It is obvious that the larger the degrees of freedom, the more precisely that variance component will be estimated. However, there are always practical (including cost) limitations on how much data can be collected. This question of assigning the degrees of freedom is now answered assuming that the total number of observations N is fixed, and that the interest centers on obtaining a precise estimate of total variance.

Suppose there are three sources of variance denoted by A, B, and C. The respective degrees of freedom are v_1, v_2, and v_3. Then,

$$v_1 + v_2 + v_3 = N-1 \tag{8.1}$$

$$s_{total}^2 = s_A^2 + s_B^2 + s_C^2 \tag{8.2}$$

The estimated sample variance has variability. If the estimated variance is based upon v degrees of freedom, then

$$\text{Variance}\,(s^2) = \frac{2\sigma^4}{v} \tag{8.3}$$

From Equations (8.2) and (8.3),

$$\text{Variance}\,(s_{total}^2) = \frac{2\sigma_A^4}{v_1} + \frac{2\sigma_B^4}{v_2} + \frac{2\sigma_C^4}{v_3} \tag{8.4}$$

Substituting for v_3 from Equation (8.1) in Equation (8.4) and then differentiating Equation (8.4) with respect to v_1 and v_2 and setting the derivatives equal to zero leads to the conclusion that the variance of s_{total}^2 is minimized when the following condition holds:

$$\frac{\sigma_A^2}{v_1} = \frac{\sigma_B^2}{v_2} = \frac{\sigma_C^2}{v_3} \tag{8.5}$$

Equation (8.5) implies that the degrees of freedom allocated to each variance component should be proportional to *our prior estimate* of the percent contribution of that variance component to the total.

Let us consider some examples. Suppose we decide to collect $N = 13$ data points. We have no idea about the relative percent contributions of the variance components. In this case, we may proceed assuming that the three variance components may be equal and give each four degrees of freedom, namely, we want to use five lots, five products, and four duplicates in an experiment where the total number of trials is 13.

Suppose based upon our previous experience, our prior estimate of the three variance components was the following: lot-to-lot $= 10$ percent, product-to-product within a lot $= 10$ percent, and duplicate $= 80$ percent. For $N = 13$, in this case there will be 1 degree of freedom allocated to both lots and products and 10 degrees of freedom allocated to duplicates. So, the experiment will involve 2 lots, 2 products, and 11 duplicates designed such that the total number of trials is 13.

8.5 QUESTIONS TO ASK

The following is particularly meant for those who work in or with manufacturing. Variance components analysis is an underutilized subject in industry. Is your company using the variance components analysis effectively? If not, you may help implement the approach by asking the following questions:

1. Based upon available data what do the distributions (histograms) of key output characteristics look like? Which characteristics require variance reduction because the out-of-specification product and economic losses are too large?

2. In the passively collected manufacturing database, what information is currently available on the "tags" (likely causes of variability) associated with these characteristics of interest?

3. If no such information is currently being collected, what information should be collected? The answer requires some speculation regarding the likely causes of variability.

4. Are the factors in the study correctly classified as fixed or random?

5. Is the structure of the study correctly defined as nested, crossed, or mixed structure?

6. Are the degrees of freedom (based upon the sample sizes) adequate to sufficiently precisely estimate each variance component?

7. What conclusions were reached based upon the variance components analysis?

8. How are variance reduction plans currently being developed? Are they based upon a clear understanding of the key causes of variability?

9. How are resource allocation decisions currently being made? Are they based upon a clear understanding of the economic consequences of variability?

   - What is the specific decision to be made? What specific improvements are sought (process centering, variance reduction, cost reduction, increased throughput, etc.)?
   - Are the costs of implementing the decision identified?
   - How are the benefits of variance reduction and process centering being quantified in economic terms?
   - How was the decision made?

Further Reading   The book by Joglekar (2003) provides a good foundation for the theory of variance components using the nested, crossed, and mixed data collection schemes. It also has several additional practical applications including those involving fixed factors and designed experiments.

CHAPTER 9

IS MY MEASUREMENT SYSTEM ACCEPTABLE AND HOW TO DESIGN, VALIDATE, AND IMPROVE IT?

Several questions are often asked in the context of a measurement system: What are the acceptance criteria for a measurement system? How to design a robust measurement system? How to demonstrate that the measurement system is acceptable? And if not, how to improve it? This chapter provides acceptance criteria for measurement precision and accuracy, for both nondestructive and destructive tests. The rationale for these acceptance criteria is explained. An example is presented to illustrate how robust product design principles can be used to design a robust measurement system. Another example shows how a designed experiment can be used to cost-effectively validate a measurement system and also to develop specifications for measurement system parameters. A gage repeatability and reproducibility (gage R&R) study shows how the acceptability of the measurement system can be assessed and how the measurement system can be improved if necessary. The design of cost-effective sampling schemes is also explained.

Measurements are often inexact. Decisions have to be made in the presence of measurement variability. Therefore, it is important to design, validate, and improve the measurement system to achieve the desired accuracy and precision.

This chapter begins by providing the rationale to establish the precision and accuracy requirements a measurement system must meet to be considered adequate. The use of robust product design ideas to design a robust measurement system is illustrated by an example. An example is presented to show how designed experiments can be used to cost-effectively validate a measurement system and also to develop specifications for

Industrial Statistics: Practical Methods and Guidance for Improved Performance   By Anand M. Joglekar
Copyright © 2010 John Wiley & Sons, Inc.

the measurement system. The very commonly used gage R&R study is presented to assess the acceptability of a measurement system and how the system can be improved if necessary. The design of cost-effective sampling schemes is explained.

9.1 ACCEPTANCE CRITERIA

All measurement systems have bias and variability. Measurement system variability includes both the repeatability and the reproducibility of the measurement system. Repeatability is measured by the standard deviation of repeated measurements of the same product under identical operating conditions. If these repeated measurements of a known standard were to be made, then the difference between the true mean measurement and the true value of the standard is bias under the fixed conditions of measurement. This bias can be estimated as the difference between the observed average measurement of the standard and the true value of the standard. As measurement conditions (operator, environment, etc.) change, the bias may change. The standard deviation of bias is a measure of reproducibility of the measurement system. Repeatability and reproducibility add as follows to constitute the total variance of the measurement system.

$$\sigma_m^2 = \sigma_{repeatability}^2 + \sigma_{reproducibility}^2 \tag{9.1a}$$

The difference between the true mean measurement of a standard (μ) under varying measurement conditions and the true value of the standard (T) is the average bias.

$$\text{Bias} = \mu - T \tag{9.1b}$$

If a large number of products from a production line are measured with respect to a certain characteristic, then the observed variability includes both the true product-to-product variance and the measurement system variance as shown in Figure 9.1. The following relationship holds.

$$\sigma_t^2 = \sigma_p^2 + \sigma_m^2 \tag{9.1c}$$

In Equation (9.1c) σ_p is the true product-to-product standard deviation, σ_m is the measurement standard deviation, and σ_t is the total standard deviation.

What values of Bias and σ_m are acceptable? The answer depends upon the purpose of taking measurements. Measurements have two key purposes: to improve products and processes (improvement purpose) and to assess whether the product is acceptable or not (specification-related purpose). The acceptance criteria for σ_m and Bias are now derived for both destructive and nondestructive measurements.

Acceptance Criteria for σ_m Let us begin by thinking about an acceptable σ_m for the purpose of improving a product. Suppose we are interested in improving the variability of a certain dimension of a product. To help make this improvement, the

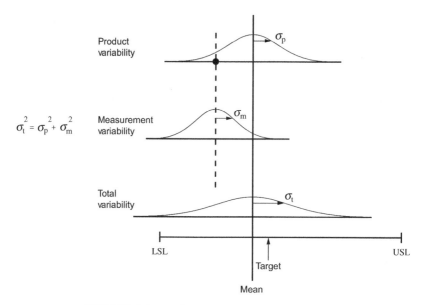

FIGURE 9.1   Product and measurement variability.

measurement system must be able to discriminate between products. It should be able to tell us that the dimensions of the two products are statistically different. Ninety-five percent of the true product dimensions will be inside $\mu \pm 2\sigma_p$ or within a total range $4\sigma_p$. Suppose we measure two products whose true measurements are μ_1 and μ_2. The measured values are x_1 and x_2. For the difference between the two true means to be detected as significant, $x_1 - x_2 \geq 2\sqrt{2}\sigma_m$. This gives us the number of distinct categories that the products can be grouped using a single measurement:

$$\text{Number of product categories} = \frac{4\sigma_p}{2\sqrt{2}\sigma_m} = \frac{1.41\sigma_p}{\sigma_m} \qquad (9.2)$$

The smaller the measurement standard deviation relative to the true product-to-product standard deviation, the greater the number of categories, as one would expect. If the number of categories is one, it means that the measurement system cannot distinguish between products. If the number of categories is two, the variable measurement system is no better than an attribute gage. If we expect the measurement system to distinguish products into at least four categories, then for improvement purposes, the acceptance criterion for σ_m is given by

$$\frac{1.41\sigma_p}{\sigma_m} \geq 4, \qquad (9.3)$$

which along with Equation (9.1) leads to $\sigma_m \leq \sigma_t/3$.

Equation (9.3) shows that the measurement standard deviation should be less than a third of the total standard deviation or that the measurement variance should be

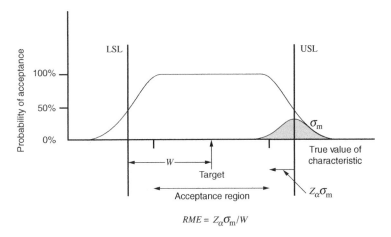

FIGURE 9.2   Relative measurement error (RME).

approximately less than 10 percent of the total variance for the measurement system to be adequate for improvement purposes.

Figure 9.2 shows the implications of measurement variability for specification-related purpose. Let W denote the half-width of a two-sided specification, or the distance between the process mean and the specification limit for a single-sided specification. It then follows that to avoid misclassification of products, as shown in Figure 9.2, a guard-banding approach will have to be taken to develop a narrower acceptance region such that the probability of accepting an out-of-specification product will be small, acceptably equal to α (say 2.5 percent). If the specification is $\pm W$, the acceptance region will become narrower equal to $\pm(W - Z_\alpha \sigma_m)$. This reduction is captured by relative measurement error (RME) defined as

$$\text{RME} = \frac{Z_\alpha \sigma_m}{W} = \frac{2.0 \sigma_m}{W} \quad \text{and for RME} = 20\%, \quad \sigma_m \leq \frac{W}{10} \qquad (9.4)$$

If $\sigma_m \leq W/10$, then approximately 20 percent of the specification will be lost to guard-banding. What the RME should be varies with the complexity of the measurement, 20 percent being a reasonable figure for moderately complex measurements. Thus, Equations (9.3) and (9.4) provide the justification for the recommended acceptance criteria in Table 9.1 for the measurement standard deviation of nondestructive measurements.

Acceptance Criteria for Bias   If we allocate half the measurement system variance to repeatability, then $\sigma_{\text{repeatability}} = 0.7 \sigma_m$. Usually, deviations in mean of the order of $1.5 \sigma_{\text{repeatability}}$ are difficult to detect with small sample sizes. This implies that Bias $\leq \sigma_m$. Thus, the acceptance criteria for Bias are the same as for σ_m.

Acceptance Criteria for Destructive Tests   Repeated measurements of the same product are not possible with destructive tests. The product-to-product

TABLE 9.1   Acceptance Criteria for σ_m and Bias for Nondestructive Tests

Purpose	σ_m (nondestructive)	Bias (nondestructive)	σ_m (destructive)	Bias (destructive)
Improvement-related	$\sigma_m \leq \dfrac{\sigma_t}{3}$	Bias $\leq \dfrac{\sigma_t}{3}$		
Specification-related	$\sigma_m \leq \dfrac{W}{10}$	Bias $\leq \dfrac{W}{10}$	$\sigma_m \leq \dfrac{W}{4}$ to $\dfrac{W}{6}$	Bias $\leq \dfrac{W}{4}$ to $\dfrac{W}{6}$

variability and $\sigma_{\text{repeatability}}$ of the measurement system are confounded with each other and cannot be separately estimated. If we want the overall C_p index in the range of 1.33–2.0, then for destructive tests

$$\sigma_m \text{ and Bias} \leq \frac{W}{4} \text{ to } \frac{W}{6} \qquad (9.5)$$

The acceptance criteria for nondestructive and destructive tests are summarized in Table 9.1. See Joglekar (2003) for further detailed discussion of the acceptance criteria.

9.2 DESIGNING COST-EFFECTIVE SAMPLING SCHEMES

Measurement system bias can be improved either by estimating the bias and adjusting for it or by redesigning or improving the measurement system. Similarly, measurement standard deviation can be reduced either by designing the measurement system to be robust or by improving the measurement system.

Situations arise where multiple measurements are feasible and the quantity of interest is the average of these multiple measurements. In these cases, appropriate sampling schemes can be designed to obtain the desired standard deviation for the measured average. These sampling schemes are often an application of variance components analysis:

1. For nondestructive tests, such as measurement of a dimension, one could take the average of multiple measurements and treat the average as the measured value. The average of n measurements will have a smaller standard deviation equal to $\sigma_m/\sqrt{n}$. The number of product categories will increase by a factor of $\sqrt{n}$.

2. In the food industry, it is common to conduct sensory tests with multiple tasters with each taster tasting the product multiple times. In this situation, if a single product is evaluated by k tasters n times then the average of these nk measurements will have a variance given below and the standard deviation of the average measurement can be reduced by increasing the number of tasters and the number of replicates.

$$\sigma_{\text{average}}^2 = \frac{\sigma_{\text{taster}}^2}{k} + \frac{\sigma_{\text{repeatability}}^2}{nk}$$

3. Similar situation occurs if we have to estimate (say) the mean drug content of a particular batch of drug material. The drug content may vary from location to location inside the batch because of particle size differences and other reasons. If we take k samples from multiple locations and test each sample n times (by taking subsamples) then

$$\sigma^2_{\text{average}} = \frac{\sigma^2_{\text{sample}}}{k} + \frac{\sigma^2_{\text{test}}}{nk}$$

As an illustration let us assume that $\sigma^2_{\text{sample}} = 4.0$ and $\sigma^2_{\text{test}} = 1.0$. Then,

- If we take one sample and test it five times, $\sigma^2_{\text{average}} = 4/1 + 1/5 = 4.2$
- If we take five samples and test each once, $\sigma^2_{\text{average}} = 4/5 + 1/5 = 1.0$
- If we take five samples and test the composite sample once $\sigma^2_{\text{average}} = 4/5 + 1/1 = 1.8$

Various alternative sampling schemes can be examined and a sampling scheme can be selected to have an acceptable value of the standard deviation of the average measurement.

4. For the above and similar situations, a cost-effective sampling scheme can be developed based upon how well we need to know the batch mean, the cost of taking a sample, cost of testing, and estimates of the variance components (Box et al., 1978).

Let S be the number of samples and T be the total number of tests performed to get the average. Then,

$$\sigma^2_{\text{average}} = \frac{\sigma^2_{\text{sample}}}{S} + \frac{\sigma^2_{\text{test}}}{T}$$

Let C_s be the cost of taking a single sample and C_t be the cost of a single test. Then, the total cost of getting an average measurement is

$$C = SC_s + TC_t$$

The ratio S/T that minimizes total cost subject to the constrain that $\sigma^2_{\text{average}}$ is constant can be obtained by differentiation and is given by:

$$\frac{S}{T} = \sqrt{\frac{\sigma^2_{\text{sample}}}{\sigma^2_{\text{test}}} \frac{C_t}{C_s}}$$

As an illustration, let us again assume that $\sigma^2_{\text{sample}} = 4.0$ and $\sigma^2_{\text{test}} = 1.0$ and that it costs \$1 to take a sample and \$10 to make one test. Then,

$$\frac{S}{T} = \sqrt{\frac{(4.0)(10)}{(1.0)(1)}} = 6.3$$

What this means is that only one test (composite sample if that is possible) should be conducted unless the number of samples exceeds six.

Various sampling schemes can be evaluated as follows keeping in mind the ratio above to get the minimum cost scheme in each case. The appropriate scheme may then be chosen based upon the desired value of the variance of the average and the cost of obtaining the average.

For $S = 1$, $T = 1$, the calculated $\sigma^2_{\text{average}} = 5.00$ and cost $= \$11$.

For $S = 3$, $T = 1$, the calculated $\sigma^2_{\text{average}} = 2.33$ and cost $= \$13$.

For $S = 6$, $T = 1$, the calculated $\sigma^2_{\text{average}} = 1.67$ and cost $= \$16$.

For $S = 9$, $T = 2$, the calculated $\sigma^2_{\text{average}} = 0.94$ and cost $= \$29$.

9.3 DESIGNING A ROBUST MEASUREMENT SYSTEM

Blood tests are done to monitor the health of people to assess kidney function, glucose levels, PSA, and blood lipids such as cholesterol, hematology, electrolytes, and so on. These tests require the development of an assay and a machine to automatically execute the assay. For the submitted blood sample, the machine measures (say) cholesterol in relative light units (RLUs), which are converted to cholesterol values using a calibration curve.

The development of the assay and the machine involves optimization of a number of potential control factors such as particle concentration, conjugate concentration, antibody input ratio, and machine processing factors. In developing a measurement system to measure cholesterol, the objective is to develop the assay and the machine such that the accuracy and precision of the measurement is the best possible over the entire range of cholesterol values that are of practical interest.

An appropriate experiment can be designed in the selected control factors as described in Chapter 3. For illustrative purposes, Table 9.2 shows a fractional factorial design in three assay control factors A, B, and C. Each trial is a prototype assay design. A calibration experiment is conducted with each prototype to cover standards ranging from 50 to 500 with replicate RLU measurements for each standard. The measurements (RLUs) are denoted by y.

Figure 9.3 shows the calibration plots for all four trials. For each trial, the calibration curve appears to be a straight line and the fitted straight line is shown in the figure. Trial 3 has the highest slope, that is, a prototype assay with a high sensitivity. However, it also seems to have the largest variability. Trial 2 has a smaller variability, but not as high a sensitivity. Of the four prototypes, which is the best?

To measure the cholesterol in an unknown sample, we first measure the RLUs for that sample, and then, use the calibration curve to predict the cholesterol level. The prototype with the smallest prediction error is the best.

TABLE 9.2   Assay Design Experiment and Data

Trial	A	B	C	50	100	200	350	500	β	σ_e	20 log (β/σ_e)
1	−1	−1	1	4.1	8.9	21.0	35.6	51.4	0.103	0.86	−18.43
				5.3	10.3	18.8	34.2	50.8			
2	1	−1	−1	4.6	7.1	16.5	27.1	38.5	0.079	0.775	−19.83
				3.7	8.2	15.6	28.8	40.2			
3	−1	1	−1	5.5	13.1	32.3	50.1	72.1	0.149	2.307	−23.80
				8.8	17.2	27.8	54.3	76.4			
4	1	1	1	4.2	14.2	26.1	40.3	61.7	0.118	2.046	−24.80
				7.7	12.2	22.6	44.3	57.4			

The linear relationship between the measurement y and the standards x is

$$y = \alpha + \beta x + e \tag{9.6}$$

where α is the intercept, β is the slope, and e is the error or residual. For an unknown sample, the value of y is measured and x is back-calculated from Equation (9.6) as

$$x = \frac{y-\alpha}{\beta} - \frac{e}{\beta} \tag{9.7}$$

(e/β) is the error made in predicting x and its standard deviation is σ_e/β, where σ_e is the standard deviation of the error. We need a prototype with the smallest prediction error standard deviation, or the largest signal-to-noise (S/N) ratio β/σ_e (Taguchi, 1987). The prototype with the largest S/N ratio is the best. The S/N ratio can be expressed in db units by the log-transformation defined below. The log-transformation

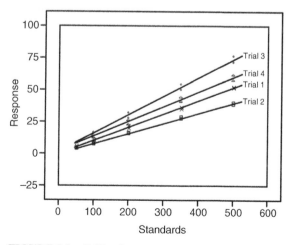

FIGURE 9.3   Calibration curves for the four prototypes.

reduce control factor interactions in the model-building stage. An improvement of 6 db units in the S/N ratio represents a reduction in prediction error standard deviation by a factor of two.

$$\text{S/N ratio for linear calibration} = 20 \log \frac{\beta}{\sigma_e} \qquad (9.8)$$

These S/N ratios are shown in Table 9.2. Prototypes 1 and 2 are superior to prototypes 3 and 4. Their S/N ratio is larger. This happens because the reduction in sensitivity for prototypes 1 and 2 is outweighed by a proportionately greater reduction in prediction error standard deviation.

We can now analyze the experiment in Table 2 with the S/N ratio as response. The analysis is done with the methods described in Chapter 3 and is also visually obvious. Factor B is the most important factor and increasing factor B reduces the S/N ratio. A formal analysis of the experiment shows that factor A has a small effect that is significant at the 90 percent confidence level. The effect of factor C is negligible. The fitted equation is:

$$20 \log \frac{\beta}{\sigma_e} = -21.7 - 0.6A - 2.6B + \text{residual} \qquad (9.9)$$

The standard deviation of residual is $\sigma_{\text{residual}} = 0.2$. The equation shows that (-1) levels of factors A and B are desirable. Going from the $(+1)$ levels of A and B to their (-1) levels increases the S/N ratio by 6.4 db. This is more than a 2:1 reduction in the standard deviation of prediction error. The results suggest that the values of A and B, particularly B, less than the (-1) levels considered in this experiment should be explored if practical.

One advantage of using the S/N ratio is that the measurement system parameters can be optimized without having to directly calculate the prediction error standard deviation for x by conducting a calibration for each trial. However, the S/N ratio given by Equation (9.8) was based on two assumptions: the calibration curve is linear and the error variance is constant throughout the calibration range. In many practical cases, these assumptions are sufficiently true. There are situations where these assumptions are violated.

For the type of data considered here, the calibration may not be linear and may be characterized by a logistic function. Also, the error variance may not be constant throughout the range of calibration. This nonconstancy of variance means that percentage errors may vary widely throughout the range of calibration if the usual unweighted regression analysis is used to obtain the calibration curve. In this case, weighted regression (Joglekar, 2003) should be performed with the logistic model. Based upon the fitted equation, the errors in predicting the standard can be obtained for each level of the standard. The CV at each level of the standard can be estimated from these prediction errors and $\log \sum CV^2$ can be obtained for each prototype trial. This should be used instead of the S/N ratio given by Equation (9.8) to build models and design the most robust measurement system by minimizing $\log \sum CV^2$.

9.4   MEASUREMENT SYSTEM VALIDATION

Measurement system needs to be validated to demonstrate its suitability for the intended use. In the pharmaceutical industry, method validation includes specificity, range, linearity, accuracy, precision, limits of detection and quantitation, and robustness (Joglekar, 2003). For example, an HPLC is robust if it is not materially affected by small variations in method parameters. These include percent acetonitrile (ACN) in the mobile phase, pH of the mobile phase, detector wavelength, column temperature, flow rate, buffer concentration, and column type (from different manufacturers). These robustness experiments are often conducted using the one-factor-at-a-time approach, which was shown to be expensive in Chapter 3. Designed experiments are a much better approach to demonstrate robustness cost effectively.

A highly fractionated fractional factorial (seven factors in eight trials) was used in this study reported in Ye et al. (2000) with capacity factor and many other responses. Interactions were expected to be small based upon prior practical experience. The ranges for each factor were selected to cover the normal expected deviations in that factor under routine operation of the HPLC. If the factors produced no practically important effect on capacity factor, then the method will be considered to be robust. Otherwise, the ranges of important factors will have to be constrained to achieve the desired capacity factor ≥ 1.5. Table 9.3 shows the designed experiment and data. There were a total of 48 trials (6 repeat injections $\times 8$ experimental conditions).

The average of all data is 1.29, that is, the average capacity factor for the entire experiment is below the desired value. The usual analysis of this experiment leads to the conclusion that all factors produce small effects except two. Percent ACN produces an effect of -1.01 and temperature has an effect of -0.52. Both these effects are large compared with the overall average. Equation (9.10) shows the relationship between capacity factor and the two important input factors in coded units. The residual standard deviation is 0.15.

$$\text{Capacity factor} = 1.29 - 0.5(\%\text{ACN}) - 0.26(\text{temperature}) + \text{residual} \qquad (9.10)$$

From Equation (9.10), it is clear that the requirement for the capacity factor to be >1.5 will not be met over the current range of percent ACN and temperature. A new specification will have to be developed for these two factors. This is done as follows and the results are shown in Figure 9.4.

1. All other factors that produce small effects were set so as to minimize capacity factor.
2. To ensure that each injection meets the requirement that capacity factor ≥ 1.5, the lower 95 percent limit on individual values of capacity factor was computed and used as a response.
3. The response surface contour for capacity factor $= 1.5$ was generated and is shown in Figure 9.4.

TABLE 9.3   Designed Experiment to Evaluate Method Robustness

Trial	%ACN	pH	Wavelength	Temperature	Flow Rate	Buffer Concentration	Column	Capacity Factor
1	13	2.8	276	35	1.7	28	2	2.18
2	13	2.8	280	45	1.3	22	2	1.60
3	13	3.2	276	45	1.3	28	1	1.32
4	13	3.2	280	35	1.7	22	1	2.21
5	17	2.8	276	45	1.7	22	1	0.63
6	17	2.8	280	35	1.3	28	1	0.76
7	17	3.2	276	35	1.3	22	2	1.24
8	17	3.2	280	45	1.7	28	2	0.72
9	13	2.8	276	35	1.7	28	2	2.17
10	13	2.8	280	45	1.3	22	2	1.59
11	13	3.2	276	45	1.3	28	1	1.38
12	13	3.2	280	35	1.7	22	1	2.21
13	17	2.8	276	45	1.7	22	1	0.60
14	17	2.8	280	35	1.3	28	1	0.76
15	17	3.2	276	35	1.3	22	2	1.22
16	17	3.2	280	45	1.7	28	2	0.71
17	13	2.8	276	35	1.7	28	2	2.17
18	13	2.8	280	45	1.3	22	2	1.59
19	13	3.2	276	45	1.3	28	1	1.39
20	13	3.2	280	35	1.7	22	1	2.21
21	17	2.8	276	45	1.7	22	1	0.60
22	17	2.8	280	35	1.3	28	1	0.76
23	17	3.2	276	35	1.3	22	2	1.22
24	17	3.2	280	45	1.7	28	2	0.71
25	13	2.8	276	35	1.7	28	2	2.17
26	13	2.8	280	45	1.3	22	2	1.59
27	13	3.2	276	45	1.3	28	1	1.38
28	13	3.2	280	35	1.7	22	1	2.22
29	17	2.8	276	45	1.7	22	1	0.60
30	17	2.8	280	35	1.3	28	1	0.76
31	17	3.2	276	35	1.3	22	2	1.21
32	17	3.2	280	45	1.7	28	2	0.71
33	13	2.8	276	35	1.7	28	2	2.17
34	13	2.8	280	45	1.3	22	2	1.58
35	13	3.2	276	45	1.3	28	1	1.34
36	13	3.2	280	35	1.7	22	1	2.22
37	17	2.8	276	45	1.7	22	1	0.60
38	17	2.8	280	35	1.3	28	1	0.76
39	17	3.2	276	35	1.3	22	2	1.20
40	17	3.2	280	45	1.7	28	2	0.71
41	13	2.8	276	35	1.7	28	2	2.16
42	13	2.8	280	45	1.3	22	2	1.58
43	13	3.2	276	45	1.3	28	1	1.43
44	13	3.2	280	35	1.7	22	1	2.22

TABLE 9.3 (*Continued*)

Trial	%ACN	pH	Wavelength	Temperature	Flow Rate	Buffer Concentration	Column	Capacity Factor
45	17	2.8	276	45	1.7	22	1	0.60
46	17	2.8	280	35	1.3	28	1	0.76
47	17	3.2	276	35	1.3	22	2	1.20
48	17	3.2	280	45	1.7	28	2	0.70

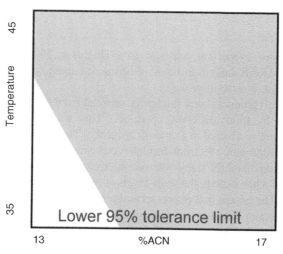

FIGURE 9.4 Determining specifications for percent ACN and temperature.

The capacity factor is greater than 1.5 to the left of the line in Figure 9.4. This means that to achieve the desired value of capacity factor, the ranges for percent ACN and temperature will have to be constrained to be in the lower left-hand corner. For example, if temperature could be controlled within 35–37 degrees and percent ACN could be controlled within 13–14 percent, the capacity factor for individual injections will meet the desired criteria. If these ranges are deemed to be too tight, then either the capacity factor requirement will have to be relaxed or ranges outside the lower left corner will have to be explored. Thus, the designed experiment leads to new specifications for measurement system parameters when validation fails.

9.5 REPEATABILITY AND REPRODUCIBILITY (R&R) STUDY

Let us now consider a typical gage repeatability and reproducibility study (gage R&R) for nondestructive tests. It uses a two variables crossed with duplicates structure to estimate product-to-product variability, operator-to-operator variability,

TABLE 9.4 Data for Gage R&R Study

Operator	Trial	Products 1	2	3	4	5	6	7	8	9	10
A	1	0.64	1.02	0.87	0.85	0.55	0.97	1.06	0.84	1.11	0.65
	2	0.60	0.99	0.80	0.95	0.49	0.91	0.98	0.89	1.09	0.60
B	1	0.52	1.05	0.78	0.82	0.37	1.03	0.87	0.75	1.02	0.77
	2	0.55	0.94	0.75	0.76	0.39	1.05	0.95	0.72	0.95	0.74
C	1	0.48	1.02	0.89	0.88	0.48	1.08	0.98	0.69	1.02	0.85
	2	0.55	1.00	0.81	0.77	0.59	1.02	0.90	0.76	1.05	0.87

product-by-operator interaction, and test error (Joglekar, 2003). The purpose of conducting a gage R&R study is to assess if the measurement system is adequate for the task, and if not, how to improve it.

For the study, 10 products were randomly selected from the production line and measured twice by three different operators. This was a blind test and the measurements were conducted in a random sequence. The collected data are shown in Table 9.4 after subtracting 99 from the observed values. In this study, products and operators are crossed and the duplicate measurements are nested. Two factors are said to be crossed when exactly the same levels of one factor are used at all levels of the other factor. The reason that operators and products are crossed in this study is that the same operator measures each product. When two factors are crossed they may interact with each other, namely, the effect of one factor may depend upon the level of the other factor. A factor is said to be nested inside another when the levels of the nested factor are all different at each level of the other factor. This is the case with duplicates.

In this study, there are four sources of variability: product-to-product variance, operator-to-operator variance, variance due to product by operator interaction, and replicate variance. These four variances add to the total variance exhibited by the collected data. The job of conducting the variance components analysis can be delegated to a software package and the results are summarized in Table 9.5. The percent contribution in Table 9.5 is each component variance expressed as a percent of the total variance.

TABLE 9.5 Variance Components from Gage R&R Study

Source	Degrees of Freedom	Variance	Standard Deviation	Contribution (%)
Product	9	0.03530	0.1880	86.3
Operator	2	0.00043	0.0208	1.1
Interaction	18	0.00320	0.0567	7.8
Test	30	0.00198	0.0445	4.8
Total	59	0.04090	0.2020	100.0

The following conclusions may be drawn from this gage R&R study.

1. The total variance (σ_t^2) is 0.04090. This partitions into product-to-product variance (σ_p^2) and measurement system variance (σ_m^2) as follows.

$$\sigma_t^2 = \sigma_p^2 + \sigma_m^2$$

   where $\sigma_p^2 = 0.03530$ and $\sigma_m^2 = 0.00562$ being the sum of the variances due to operators, product-by-operator interaction, and test method. Thus, approximately 86 percent of the total variance is due to product-to-product differences and the remaining 14 percent is due to the measurement system. Note that it is the individual variances that add to the total variance. Standard deviations do not add to the total standard deviation.

2. Let us suppose that the specification for the measured characteristic is 100 ± 0.5. Since the data in Table 9.4 are reported after subtracting 99, the specification becomes 1.0 ± 0.5. Is the measurement system adequate? From Table 9.1, for improvement purposes, the acceptance criterion is $\sigma_m < \sigma_t/3$. Presently, $\sigma_m = 0.075$ and $\sigma_t = 0.202$ so that to be acceptable σ_m needs to be 0.067. For specification-related purposes, the criterion is $\sigma_m < W/10$. Presently, $W = 0.5$ and to be acceptable σ_m needs to be 0.05. Hence, the conclusion is that the measurement system is not acceptable for improvement and specification-related purposes.

3. The total measurement variance can be further partitioned into variances due to repeatability and reproducibility, and from Table 9.5

$$\sigma_m^2 = \sigma_{repeatability}^2 + \sigma_{reproducibility}^2$$

$$\sigma_{repeatability}^2 = 0.00198$$

$$\sigma_{reproducibility}^2 = 0.00043 + 0.00320 = 0.00363$$

   The variance due to repeatability is the same as the test variance. The variance due to reproducibility is the sum of the operator and interaction variances, both these components being due to the differences between operators, and will disappear if a single operator is used to take measurements or if all operators measure alike. Repeatability accounts for 4.8 percent of the total variance and reproducibility accounts for 8.9 percent of the total variance. These add to the measurement system variance, which is 13.7 percent of the total variance.

4. Variance due to reproducibility is further divided into consistent and inconsistent differences between operators. Consistent difference between operators measures the variability due to the true differences between the average measurements of each operator. In Table 9.5, the source of consistent

differences is referred to as operator. If the differences between operators are not always consistent and depend upon the product being measured, then a product-by-operator interaction exists, referred to in Table 9.5 as interaction. For example, if for some products operator A measurements are higher than operator B measurements, but for other products the reverse is true, then this inconsistency signifies interaction. From Table 9.5, the consistent differences between operators account for 1.1 percent of the total variance. The inconsistent differences account for 7.8 percent of the total variance.

5. If by providing further training regarding measurement procedures, by providing appropriate jigs and fixtures, and so on, the consistent and inconsistent operator differences could be eliminated, then σ_m will become equal to 0.0445, the test–retest standard deviation. The reason for partitioning reproducibility into consistent and inconsistent differences is to help identify corrective actions. Generally, consistent operator differences are easier to correct than inconsistent differences.

   If σ_m reduces to 0.0445, the new σ_t will be 0.1931. W will continue to be 0.5. Now $\sigma_m < \sigma_t/3$ and the measurement system will work well for improvement purposes. Also, the new $\sigma_m < W/10$ and the measurement system will be satisfactory for specification-related purposes. Further optimization of the measurement system will not be necessary to reduce $\sigma_{\text{repeatability}}$.

9.6 QUESTIONS TO ASK

Measurement system design and analysis involve the use of many of the statistical methods described in this book. These include design of experiments, robust design, comparative experiments, variance components analysis, model building, and statistical process control. Thus, the questions to ask described in those chapters apply and are not repeated here in their entirety.

9.6.1 Measurement System Design

1. Are the objectives of the measurement system (responses to be measured, measurement cost, and cycle time) clearly defined?
2. Have the measurement system noise factors and their ranges been identified? Is it possible to create a joint noise factor, or a small set of noise conditions that capture the range of noise likely to be encountered in practice? Are facilities available to be able to experiment with noise?
3. Are the potentially important control factors and their ranges identified?
4. Has the appropriate S/N ratio been defined?
5. Is the robust design experiment structured considering the likely control factor interactions, curvature effects, and control by noise interactions?
6. What are the conclusions from the experiment and what are the next steps?

9.6.2 Measurement System Validation and Improvement

1. What are the specific purposes of the measurement study under consideration? These studies include calibration study, stability study, gage R&R study, robustness study, intermediate precision study, linearity study, method transfer study, and so on. (see Joglekar (2003) for further details).
2. Can several of the studies be combined to reduce the overall effort?
3. For the study under consideration, are the acceptance criteria clearly defined as a part of the study protocol?
4. Is the study designed with adequate sample sizes?
5. For the study under consideration, does the protocol identify the preferred method of analysis? These include weighed and unweighted regression for calibration and linearity studies, control charts (or autocorrelation function) for stability study, variance components analysis for gage R&R study, analysis of a designed experiment for robustness and intermediate precision studies, and confidence interval analysis for method transfer study.
6. What conclusions were reached and what are the next steps?

Further Reading   The book by Joglekar (2003) provides the rationale for the acceptance criteria in detail. It also discusses many different types of measurement system studies including calibration studies, stability and bias studies, linearity studies, and method transfer studies. The book by Taguchi (1987) develops the signal-to-noise ratios for the design of measurement systems and provides several examples. The book by Box et al. (1978) has a brief section on variance components and sampling schemes.

CHAPTER 10

HOW TO USE THEORY EFFECTIVELY?

While technical professionals learn a great deal of theory during their undergraduate and graduate education, theory is often not extensively and effectively used, perhaps because it is felt that theory does not work perfectly in practice. There is much to be gained, however, by the judicious use of theory coupled with data to formulate mathematical models. The purpose of this chapter is to introduce the subject of model building, both empirical modeling based purely upon the data and mechanistic modeling based upon an understanding of the underlying phenomenon.

The previous chapters demonstrate how an understanding of the relationship between inputs and outputs is useful to design products and processes, to increase the robustness of products and processes, and to set functional specifications. This relationship is often called a model. The purpose of this chapter is to further discuss the subject of model building. Two approaches to model building are considered: the empirical approach and the mechanistic approach.

Empirical models are relationships between inputs and outputs derived purely based upon data. If data are collected on the hardness of rubber and the abrasion loss of rubber, a plot of the data may suggest that as the hardness increases, abrasion loss reduces. Over the range of hardness of interest, the plot may look like a straight line with a negative slope. Regression analysis could be used to describe abrasion loss as a linear function of hardness. These empirical models are valid over the ranges of input factors. Their ability to predict outside the experimental zone is limited. Empirical

Industrial Statistics: Practical Methods and Guidance for Improved Performance   By Anand M. Joglekar
Copyright © 2010 John Wiley & Sons, Inc.

models are adequate for many purposes, including for the purpose of product and process optimization and specification development. These models are not necessarily based upon a theoretical understanding of the subject matter. Attempting to understand the reasons behind the empirically derived equations can lead to mechanistic models and scientific and technological breakthroughs.

Mechanistic models are based upon an understanding of the underlying mechanism of action. In the case of abrasion loss and hardness, this will involve understanding how abrasion of rubber occurs based upon the structural properties of rubber, the forces acting on it, and so on. This understanding may suggest an exponential or power function relationship, the observed straight line being a reasonable approximation to the mechanistic relationship within a narrow range of interest. The derivation of these theoretical equations is a common practice in some industries—electronic circuit design, fluid flow modeling, and design of mechanical systems being some examples. If a mechanistic model was available, such as for a circuit design problem, then the selection of optimum targets and tolerances for all circuit components can be done by pure simulation without having to do any real experiments. However, obtaining these adequate mechanistic models is not easy. Therefore, judgment is needed in deciding when the effort to develop a mechanistic model may be justified. This attempt is justified when (1) the cost of conducting an experiment to obtain an empirical model adequate for the task is very large, (2) when a basic theoretical understanding of the subject is essential to making progress, as would often be the case in early research, and (3) when it is relatively easy to obtain these models because the necessary theory is well understood. Mechanistic models contribute to scientific understanding, are valid over wider ranges, and are likely to involve fewer parameters that an empirical model. The attempt to derive mechanistic models, even if unsuccessful, is very useful to making progress. This is especially true in the research phase of a project.

While people learn a great deal of theory in their undergraduate and graduate education, in many industries, they do not seem to use it to the fullest extent possible. Perhaps this is because of their experience that the known theory does not apply sufficiently well in practice. However, very modest efforts to write down a likely equation and examine its implications can prove very useful. Consider the following simple situation. An automatic pipette is used to collect a certain volume of drug-containing sample and the sample is analyzed to determine the drug content. Instead of simply saying that the drug content depends upon volume and concentration, taking the small next step to write the following equation is useful.

$$\text{Drug content} = \text{volume} \times \text{concentration}$$

This equation suggests that, as concentration changes, the standard deviation of drug content will change proportional to the concentration. It is the CV of the drug content that is likely to stay constant. If a calibration experiment were to be conducted with different concentration standards, weighted regression should be used. Had the equation involved unknown constants to be experimentally estimated, conducting designed experiments in the log metric would have been appropriate. Just knowing whether the likely form of the equation is additive or multiplicative is very useful. We will see more examples a little later.

10.1   EMPIRICAL MODELS

Some drug-containing tablets are spray-coated in a pan coater with materials that control drug release. To determine the effect of coating time on coat weight CV, three coating runs were conducted with a pan coater. At coating time = 1, 3, 5, and 10 hours, 100 tablets each were withdrawn and their coat weights were measured. A plot of the calculated coat weight percent CV versus coating time is shown in Figure 10.1(a). The CV of coat weight is large at low coating times and reduces as coating time increases.

We have to express CV as a function of time. What is the likely form of this relationship based upon Figure 10.1(a)? Once the form of the equation is specified, the task of estimating the parameters of the equation can be delegated to a regression analysis software package. Clearly, the relationship is not linear. It seems more like an exponential. Also, the variability in CV seems to be larger for larger values of CV, which occur at lower coating times. These observations suggest that the relationship between log CV and log time may be linear. This is indeed the case as shown in

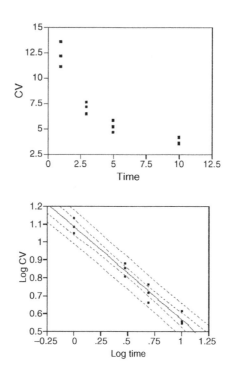

FIGURE 10.1   (a) Coat weight CV versus coating time and (b) log CV versus log time.

TABLE 10.1   Estimated Parameters and Confidence Intervals

Term	Estimate	Standard Error	95% Confidence Interval
Intercept	1.09	0.02	1.045 to 1.135
Slope	−0.52	0.03	−0.587 to −0.453

Figure 10.1(b). In the log–log metric, the relationship appears to be linear and the variability of log CV appears to be constant over time.

Using the standard regression analysis approach explained later, the fitted equation for the straight line is:

$$\text{Predicted log CV} = 1.09 - 0.52 \log \text{time} \qquad (10.1a)$$

Equation (10.1a) can be rewritten as:

$$\text{Predicted CV} = k(\text{time})^{-0.52} \quad \text{where} \quad k = 10^{1.09} = 12.2$$

It turns out that the 95 percent confidence interval for slope ranges from −0.587 to −0.453 and includes the value −0.5 as shown in Table 10.1. If the slope is taken to be −0.5, then the equation simplifies to:

$$\text{Predicted CV} = \frac{k}{\sqrt{\text{time}}} \qquad (10.1b)$$

The CV reduces inversely proportional to the square root of coating time. If the coating time is increased by a factor of four, the CV reduces by a factor of two. This is the empirical understanding gained by analyzing the collected data. For fixed coating conditions, the relationship allows us to tradeoff the cost of coating for a longer duration of time versus the benefits of the anticipated reduction in coat weight CV. Analyzing the scientific reasons for this relationship could lead to a mechanistic understanding of the coating process.

Statistics of Regression   The basic steps in fitting an equation to data (also known as regression analysis) are briefly explained below. For a detailed treatment of the subject, see Neter et al. (1990).

1. The first step is to plot the data as a scatter diagram (Figure 10.1(a)) and decide the equation to be fitted. Further recommendations on selecting equations follow in the next section. Sometimes, the relationship can be linearized, as was the case with coat weight CV.
2. With reference to Figure 10.1(b), we need to fit a straight line given by

$$y_i = a + bx_i + e_i$$

The task of estimating the intercept a and slope b is usually delegated to a computer program. The parameters are estimated to minimize the residual sum of squares $\sum e_i^2$. The estimated values of the parameters and their confidence intervals are shown in Table 10.1. A number of software packages do not calculate the confidence intervals for the intercept and the slope. These can be calculated as follows:

$$\text{Confidence interval for } a\text{: } a \pm t_{1-\alpha/2,n-2}s(a)$$

$$\text{Confidence interval for } b\text{: } b \pm t_{1-\alpha/2,n-2}s(b)$$

where $s(a)$ and $s(b)$ are the standard deviations (also called standard errors) of the estimated parameters, which are usually provided by the computer programs.

From Table 10.1, the confidence intervals do not include zero, that is, both the parameters are statistically significant. A slope of -0.5 is within the confidence interval for slope, suggesting the $\sqrt{\text{time}}$ relationship discussed earlier.

In Figure 10.1(b), the inner two dotted lines are the confidence intervals on the fitted regression line, that is, we are 95 percent sure that the true relationship is inside these two dotted lines. The outer two dotted lines are the tolerance intervals on the individual data values, that is, we are 95 percent sure that any future observations will be inside these two outer lines. These sets of lines, which are hourglass-shaped because we know the middle portion of the line more precisely than the edges, provide a feel for how precise the fitted line is and how large the deviations from the fitted line could be.

3. The adequacy of the fitted model is judged in several ways. R^2 is the square of the correlation coefficient between the input and the output and measures the percentage of variability of output explained by the input. In this case, $R^2 = 0.97$, that is, 97 percent of the variance of output is explained by the fitted equation. The portion of output that is not explained by the model is denoted by the residual e and the variance of this unexplained portion accounts for the balance of 3 percent of the variance of output. We need the R^2 to be high while ensuring that every parameter of the fitted equation is statistically significant. From Figure 10.1(b), R^2 can be increased simply by fitting a cubic equation to these data instead of a straight line. But in this case, the quadratic and cubic coefficients turn out to be statistically insignificant and will have to be dropped from the model. The idea is to only retain statistically significant terms in the equation, and then, examine R^2. If R^2 is low, it would suggest that an important factor is not included in the equation or that the measurement errors are very large.

More important than R^2 is the prediction error associated with the model. The standard deviation of prediction error in this case is 0.04, that is, if this equation is used to make predictions, the predicted log CV will have a 95 percent confidence interval of approximately ± 0.08. If the predicted log CV is 1.0, the true value of log CV will be in the interval of 0.92–1.08, from which the interval for predicted CV can be computed. If log $CV = 1.0$, then $CV = 10$ and the 95 percent confidence interval is 8.3–12.0.

4. The model can be further diagnostically checked by conducting an analysis of the residuals. The residuals are assumed to be independent, normally distributed with zero mean and constant variance. This diagnostic checking is done by plotting the residuals against variables in the model, fitted values, variables not included in the model but for whom data may have been collected, the sequence of data collection, and so on. These plots may indicate how the model can be further improved.

Transformations   What if the relationship between the output y and the input x as shown by a scatter plot is curvilinear? How can one guess at the likely model? A knowledge of the shapes that simple equations can take proves useful in this regard.

Suppose the scatter plot looks like an exponential decay. It is then likely that the relationship may be:

$$y = ae^{bx} + \text{residual}_y \tag{10.2a}$$

A model is said to be linear in parameters if the derivative of y with respect to the parameters is not a function of the parameters. Otherwise, the model is nonlinear in parameters. Parameters of nonlinear models have to be iteratively estimated using nonlinear regression. The model in Equation (10.2a) is nonlinear. It can be made linear by taking natural logarithm.

$$\ln y = \ln a + bx + \text{residual}_{\ln y} \tag{10.2b}$$

The coefficients a and b can now be estimated by linear regression. Should we use the nonlinear model (10.2a) or the linear model (10.2b)? The answer depends upon which model residuals are normally, independently distributed with zero mean and constant variance. Nonconstancy of variance and nonnormality of residuals frequently go together. If residual_y has a nonnormal distribution or if the variance of residual_y increases as y increases, then we need to transform y since the shape and spread of the distribution of y need to be changed. It often happens that the necessary transformation of y also helps linearize the equation. In this case, Equation (10.2b) is often the right approach where $\text{residual}_{\ln y}$ turns out to have the desired properties. What if residual_y has a normal distribution with constant variance? Then, nonlinear regression should be used or a transformation on x should be attempted, by transforming it to $x' = e^{bx}$ with some known values of b that make practical sense in the context of the problem at hand, such as -1 and 1. If this value of b is sensible, then Equation (10.2a) can be written as the linear Equation (10.2c), which is linear in x'.

$$y = ax' + \text{residual}_y \tag{10.2c}$$

Table 10.2 shows some useful two-parameter non-linear equations and their linear forms that could be attempted always ensuring that the residuals of the fitted model are normally, independently distributed with zero mean and constant variance. Otherwise,

TABLE 10.2   Useful Two-Parameter Equations and Their Linear Forms

Function	Equation	Linear Form
Power	$y = ax^b$	$\ln y = \ln a + b(\ln x)$
Logarithmic	$y = a + b(\ln x)$	$y = a + b(\ln x)$
Exponential	$y = ae^{bx}$	$\ln y = \ln a + bx$
Hyperbolic	$y = \dfrac{x}{(ax-b)}$	$\dfrac{1}{y} = a - b\left(\dfrac{1}{x}\right)$
Special	$y = ae^{b/x}$	$\ln y = \ln a + b\left(\dfrac{1}{x}\right)$
Special	$y = \dfrac{1}{(a + be^{-x})}$	$\dfrac{1}{y} = a + b(e^{-x})$

equations with a larger number of parameters or equations nonlinear in parameters, which require iterative parameter estimation, may be necessary.

The Box–Cox transformations discussed in Chapter 3 should also be considered to correct for nonnormality and unequal variance. These transformations are in the family of power transformations of y of the type y^{λ}. This family includes the simple transformations y^2, $\sqrt{y}$, $\ln y$, $1/\sqrt{y}$, and $1/y$. Many a time, these transformations also result in linear relationships. Sometime, they can induce nonlinearity requiring a transformation on x. To summarize,

1. The purpose of empirical modeling is to capture the relationship between the inputs and outputs in a form that is satisfactory for the practical purposes at hand.

2. The form of the relationship is often unknown and is guessed by looking at the scatter plot and by knowing the possible shapes produced by different equations. Some trial transformations of y and x, intended to produce a linear relationship in the transformed y and x, are helpful in this process.

3. The residuals of the fitted equation should be normally, independently distributed with zero mean and constant variance. Otherwise, weighted regression should be used.

4. It is most satisfying and conducive to further thinking if the fitted model makes physical sense and the coefficients have practical, useful interpretations. If the scatter diagram between the price of a commodity and quantity demanded looked similar to Figure 10.1(a), then economists will prefer a logarithmic transformation on both price and quantity because then the slope of the regression line in the transformed variables can be interpreted as price elasticity of demand, namely, the percentage change in quantity demanded per percent change in price.

5. Many times, in designed experiments, the response is a curve. For example, in conducting experiments with drug formulations, the response may be a release rate profile over time. One approach to analyze this response is to discretize it at

several time points and then analyze the response at each time point separately. Another approach is to use a piecewise linear approximation for the curve. A more satisfactory approach is to fit an equation to the curve and analyze the parameters of the equation as responses.

10.2   MECHANISTIC MODELS

Let us now consider three different mechanistic models to illustrate how these models may be obtained and their many advantages. The injection force example illustrates how previously available theory can be slightly modified to be useful in a new application, and how this model reduces the product development effort dramatically. The accelerated stability example illustrates how mechanistic models permit greater extrapolation and thereby dramatically reduce the test time. The coat weight CV example shows how useful mechanistic models can be constructed from first principles and how they not only lead to a greater understanding of the system, but also reduce the system optimization effort.

Injection Force Example   It is important to have a reasonable injection force to administer medication to a patient. Given a fluid of certain viscosity, injection force (F) is known to be a function of four factors: needle diameter (r), needle length (L), syringe diameter (R), and injection speed (Q). For a Newtonian fluid, the theoretical equation or mechanistic model describing the relationship is:

$$F \propto \frac{r^2 L Q}{R^4} \tag{10.3}$$

The fluid under consideration was a non-Newtonian fluid for which the force equation was unknown. A reasonable approach seemed to be to generalize the above equation as a power function, and then, to estimate the coefficients experimentally. This approach may be called an *empirical mechanistic* approach in that the form of the equation is mechanistic, but the parameters of the equation are fitted empirically.

$$F = k \frac{r^a L^b Q^c}{R^d} \tag{10.4}$$

The nonlinear equation can be rewritten as:

$$\log F = \log k + a \log r + b \log L + c \log Q - d \log R \tag{10.5}$$

This is a linear equation *in log terms* and, if required, the coefficients ($k, a, b, c,$ and d) can be estimated in as few as five trials.

If the necessary theoretical thinking had not been done, it is possible that the situation could have unfolded as follows using the standard design of experiments approach. Note that any equation, such as Equation (10.4), can be expanded as an infinite Taylor series expansion. This infinite series can be truncated after the quadratic

TABLE 10.3   Injection Force Experiment

Trial	Needle ID	Needle Length	Syringe ID	Injection Speed	Force
1	0.2	12	2	0.05	15
2	0.3	50	3	0.5	170
3	0.3	12	3	0.05	9
4	0.2	12	3	0.5	176
5	0.3	50	2	0.05	14
6	0.3	12	2	0.5	32
7	0.2	50	3	0.05	124
8	0.2	50	2	0.5	155
9	0.25	25	2	0.25	48

terms to obtain an approximate quadratic equation that will hopefully be satisfactory in the region of experimentation. Had we followed the usual screening followed by optimization approach to design of experiments (Chapter 3) with the only knowledge that there were four potential control factors, then it would have taken us about 25 trials to obtain this approximate quadratic equation. Instead, we are now in a position to fit an equation that makes theoretical sense and is applicable over a wide range with as few as five and no more than eight trials.

An eight-trial fractional factorial design with a near center point was used as shown in Table 10.3. While only five trials are necessary to estimate the model parameters, the additional trials permit an assessment of the adequacy (departure from the assumed form of the equation) of the model. The analysis was done after log-transforming the response and the control factor levels. The interaction and quadratic terms turned out to be unimportant, as expected from Equation (10.5). This suggests that the assumed form of the equation is adequate. By reverse transformation, the fitted equation was:

$$F = 0.07 \frac{r^{2.08} L^{0.72} Q^{0.7}}{R^{2.64}} \tag{10.6}$$

This equation was used to select values of control factors necessary to obtain satisfactory force performance.

Based upon Equation (10.6), it can be shown that

$$CV^2(F) = 2.08^2 CV^2(r) + 0.72^2 CV^2(L) + 0.7^2 CV^2(Q) + 2.64^2 CV^2(R) \tag{10.7}$$

Equation (10.7) allows us to examine the variability in force as a function of variability in needle ID, needle length, injection speed, and syringe ID. If a large variability in force is undesirable, then Equations (10.6) and (10.7) can be used to set specifications for the four control factors.

Accelerated Stability Tests   This application of regression analysis to accelerated battery stability tests is taken from Joglekar (2003). A similar approach may be

TABLE 10.4 Accelerated Battery Stability Test

Temperature T (K)	Time t (months)	Resistance y_t (ohms)	$\ln\left[\dfrac{\ln(y_t/y_0)}{t}\right]$	$1/T$	Sample Size
298	6	13.27	0.069599	0.003356	12
298	9	21.73	0.101198	0.003356	12
298	12	36.3	0.118659	0.003356	12
313	3	20.52	0.284497	0.003195	12
313	6	62.98	0.329151	0.003195	12
313	9	115.15	0.286481	0.003195	12
323	1	15.47	0.570992	0.003096	12
323	2	35.03	0.694147	0.003096	12
323	2.5	44.58	0.651750	0.003096	12
323	3	70.07	0.693862	0.003096	12

used to test drug stability or the shelf life of a food product. The response of interest was the internal resistance of the battery, which increased upon storage. The specification required the internal resistance to be less than 500 ohms over 2-year storage at room temperature (25 °C). The purpose of conducting accelerated stability tests was to establish acceleration factors so that appropriate short duration accelerated tests could be conducted to evaluate battery performance.

Data from the accelerated stability tests are shown in Table 10.4. The initial resistance of a large number of batteries was measured and the average was found to be 9 ohms. Batteries were stored at three temperatures: 25 °C, 40 °C, and 50 °C that respectively translate to 298 K, 313 K, and 323 K. Twelve batteries were periodically withdrawn and their observed average resistance values are reported in Table 10.4. It was anticipated that the change in resistance would be described by a first-order reaction.

Mechanistic models are often formulated in terms of differential equations. For a first-order reaction, the rate of change of resistance is proportional to the value of the resistance.

$$\frac{dy_t}{dt} = k_1 y_t$$

where y_t is the battery resistance at time t. Upon integration,

$$\ln y_t = \ln y_0 + k_1 t$$

where y_0 is the initial resistance of 9 ohms. The rate constant k_1 increases with temperature and it was assumed that the relationship between k_1 and absolute temperature can be expressed by the following Arrhenius equation:

$$k_1 = Ae^{-(E/RT)}$$

where T is the absolute temperature in degrees Kelvin, E is the activation energy, R is the gas constant, and A is a constant. Using the Arrhenius equation for k_1

$$\ln y_t = \ln y_0 + Ae^{-(E/RT)}t$$

Rearranging the terms, we get the following equation for a straight line:

$$\ln\left[\frac{\ln(y_t/y_0)}{t}\right] = \ln A - \frac{E}{RT} \tag{10.8}$$

Fitting Equation (10.8) to the data presented in Table 10.4 leads to

$$\ln\left[\frac{\ln(y_t/y_0)}{t}\right] = 22.57 - \frac{7432}{T} \tag{10.9}$$

The model provides a good fit to the data with $R^2 = 97$ percent.

Equation (10.9) can now be used to establish the acceleration factors necessary to be able to predict battery life at room temperature based upon accelerated tests conducted at higher temperatures. If t_{25} and t_{50}, respectively, denote the time to reach any specific resistance of y^* ohms at 25 °C and 50 °C, then the acceleration factor (AF) may be defined as

$$AF = \frac{t_{25}}{t_{50}}$$

Using Equation (10.9), we get the following two equations at 25 °C (298 K) and 50 °C (323 K).

$$\ln\ln(y^*/9) - \ln t_{25} = 22.57 - \frac{7432}{298}$$

$$\ln\ln(y^*/9) - \ln t_{50} = 22.57 - \frac{7432}{323}$$

By subtracting the first equation from the second,

$$\ln\left(\frac{t_{25}}{t_{50}}\right) = \frac{7432}{298} - \frac{7432}{323} = 1.93$$

Hence,

$$AF = \frac{t_{25}}{t_{50}} = e^{1.93} = 6.9$$

This means that 50 °C accelerates the degradation rate by 6.9 times compared with 25 °C so that 2-year storage at 25 °C is equivalent to 3.5-month storage at 50 °C. In this manner, accelerated stability tests permit the prediction of shelf life with small duration tests. These predictions would have been difficult without the mechanistic models provided by the first-order reaction and the Arrhenius equation.

10.3 MECHANISTIC MODEL FOR COAT WEIGHT CV

Let us now consider an example of how a mechanistic model may be derived from first principles. Pan coating processes have been used in the pharmaceutical industry for many years. Film coatings are applied to solid dosage forms such as a controlled release tablet. The drug release is a function of coat weight and the permeability of coating. Large variability in drug release is undesirable for the patients and also for the producer because it can cause lot failures in manufacturing. This implies the need for small variability of coat weight. Understanding how various coating factors control coat weight variability is important. Large-scale designed experiments are very expensive to conduct, and this was the driver for attempting to obtain a mechanistic model for coat weight CV (see Joglekar et al. (2007)).

As shown in Figure 10.2, in a typical pan coating process, a bed of tablets and a spray nozzle are placed inside a rotating drum. As the drum rotates, a cascading layer of tablets is formed on the surface of the bed. When the tablets in the cascading layer enter into the spray zone, they are coated by an atomized solution. The tablets then move back into the bulk of the tablet bed where evaporation of the solvent is aided by the warm air flowing through the bed. At some point in time, the same tablet reenters the spray zone and this process is repeated.

Derivation of the Model   The proposed mechanistic model is derived by breaking down the continuous coating process into a large number of discrete but similar coating events. A key assumption made in deriving the equation is that the mixing of tablets is completely random. The following nomenclature is used.

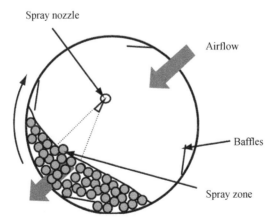

FIGURE 10.2   Schematic diagram of a pan coater.

J = number of spray guns

R = spray rate per gun (mg per minute)

t = total coat time (minutes)

N = total number of tablets being coated

a = average projected area of the tablet (cm^2)

V = velocity of the tablet in the spray zone (cm per minute)—related to rotational velocity

L = length of the spray zone (cm)—dimension of a rectangular spray zone perpendicular to the direction of tablet travel

M = width of the spray area through which the tablet travels (cm)

A = rectangular spray area = LM (cm^2)

Several key assumptions are made in deriving the model and these are stated as the model derivation proceeds.

Assumption 1: The coater has J spray guns with nonoverlapping spray areas. Each gun has the same spray area A, considered to be a rectangular coating zone of length L and width M through which the tablet travels. Each tablet has the same projected area a on which coating gets applied in the coating zone.

Assumption 2: Once the tablet gets in the coating zone, it stays there for a fixed period of time Δt called the coating event, where a coat weight Δw is applied to it. Each coating event has ΔN tablets in it.

The coat weight applied to the tablet per coating event is given by

$$\Delta w = \frac{JR\,\Delta t}{\Delta N} \tag{10.10}$$

Assumption 3: The mixing of the tablets is assumed to be perfectly random, namely, each tablet has an equal probability p of being in the coating event. It follows that

$$p = \frac{\Delta N}{N} \tag{10.11}$$

$$\Delta N = \frac{JA}{a} \tag{10.12}$$

The number of coating events k during the entire coating run is given by

$$k = \frac{t}{\Delta t} \tag{10.13}$$

The number of coating events per tablet during the entire coating run, denoted by x, has a binomial distribution with parameters k and p, where k is the total number of coating events during the entire coating run and p is the probability of a specific tablet participating in a specific coating event.

$$\text{Probability}(x) = \frac{k!}{(k-x)!(x)!}p^x(1-p)^{k-x} \tag{10.14}$$

From Equations (10.11) and (10.13)

$$kp = \frac{t\,\Delta N}{N\,\Delta t} \tag{10.15}$$

Based upon the properties of the binomial distribution, the mean, standard deviation, and CV of the number of coating events per tablet x, respectively denoted by μ_x, σ_x, and CV_x are given by

$$\mu_x = kp = \frac{t\,\Delta N}{N\,\Delta t}$$

$$\sigma_x = \sqrt{kp} = \sqrt{\frac{t\,\Delta N}{N\,\Delta t}} \tag{10.16}$$

$$CV_x = \frac{100}{\sqrt{kp}} = 100\sqrt{\frac{N\,\Delta t}{t\,\Delta N}}$$

Assumption 4: Each tablet travels across the width M of the coating zone at velocity V (related to the rotational velocity of the drum).

$$\therefore \quad \Delta t = \frac{M}{V} \tag{10.17}$$

Using Equations (10.10), (10.12), and (10.17) and noting that $A = LM$,

$$\Delta w = \frac{aR}{LV} \tag{10.18}$$

Using Equations (10.15), (10.12), and (10.17) and noting that $A = LM$,

$$kp = \frac{tJVL}{aN} \tag{10.19}$$

Coat weight per tablet is Δw multiplied by x. The mean, standard deviation, and CV of x are given by Equation (10.16). Multiplying Equation (10.16) by Δw and substituting for Δw and kp using Equations (10.18) and (10.19) gives us the mean, standard deviation, and CV of coat weight per tablet, respectively denoted by μ_{cw}, σ_{cw}, and CV_{cw}.

$$\mu_{cw} = \frac{JRt}{N} \tag{10.20}$$

$$\sigma_{cw} = R\sqrt{\frac{atJ}{VLN}} = \sqrt{\frac{aR\mu_{cw}}{VL}} \tag{10.21}$$

$$CV_{cw}\,(\%) = 100\sqrt{\frac{aN}{VLJt}} = 100\sqrt{\frac{aR}{VL\mu_{cw}}} \tag{10.22}$$

For the details of an experimental verification of this model, see Joglekar et al. (2007).

An Example   As an example of the use of the model, given by Equation (10.19), consider a coating run with coating parameters given below:

$J = 4$

$R = 8125\text{ mg/minute}$

$t = 812\text{ minutes}$

$N = 600{,}000\text{ tablets}$

$V = 4000\text{ cm/minute}$

$A = 6\text{ inches by }1.75\text{ inches} = 68\text{ cm}^2$

$M = 1.75\text{ inches} = 4.45\text{ cm}$

$L = 6\text{ inches} = 15.24\text{ cm}$

$a = 0.32\text{ cm}^2$

Substituting these parameters in the various equations, we obtain the following results:

1. $\Delta N = 850$. This means that on average 850 tablets participate in a coating event.
2. $\Delta t = 0.0011\text{ minutes} = 0.066\text{ seconds}$. This is the length of time taken by a tablet to cross the coating zone. This is the coating event time.
3. $\Delta w = 0.0425\text{ mg}$. On the average, this is the amount of coat weight applied to each tablet in a coating event.
4. $k = 730{,}000$. This is the number of coating events that occur during the entire coating run.
5. $p = 0.00142$. This is the probability that a random tablet will participate in a randomly selected coating event.
6. $\mu_x = 1034$. This is the average number of coating events that a tablet participates in during the entire coating run.
7. $\sigma_x = 32$. This is the standard deviation of the number of coating events that a tablet participates in. A tablet on the average participates in a 1000 coating events with the 3-sigma limits from 900 to 1100 coating events.

8. $CV_x = 3.1$ percent. This is the coefficient of variation of the number of coating events per tablet.

9. $\mu_{cw} = 44$ mg. This is the average coat weight applied to the tablet.

10. $CV_{cw} = 3.1$ percent. This is the coat weight CV under the assumption of random mixing.

Implications of the Model   The mechanistic model helps scale-up the coater and also shows how the coater can be optimized. The major conclusions that can be drawn from the model given by Equations (10.20) and (10.22) are described below.

1. A scale-up factor is suggested by Equation (10.22) to keep CV constant during scale-up and corresponds to the average number of coating events per tablet given by Equation (10.19).

$$\text{Scale-up factor} = \frac{VLJt}{aN}$$

If several different pan coaters using varying coating conditions have the same value for the scale-up factor, their coat weight CVs are expected to be the same. During scale-up, the intent is to scale each coating event and then keep the average number of coating events per tablet constant.

2. For a fixed desired coat weight, the CV of coat weight can be reduced by reducing the tablet size, reducing the spray rate (which implies increasing the coating time), increasing the rotational velocity of the drum, and eliminating any gaps in the spray zone without adding overlap.

3. If we double the number of tablets and double the coating time, the mean and CV of coat weight are unchanged.

4. If we double the number of tablets and double the spray rate to keep the total coating time the same, the mean coat weight will remain the same, but the CV will increase by a factor of $\sqrt{2}$.

5. If we increase the coat time and reduce the spray rate to keep the mean coat weight the same, CV will reduce proportional to $\sqrt{\text{time}}$.

6. CV is proportional to the square root of the tablet-projected area, or roughly proportional to the tablet diameter (for a spherical tablet).

7. If the observed CV is much larger than the predicted CV, then it is likely that the mixing of tablets is not perfectly random suggesting the need to improve mixing through improved baffle design.

10.4   QUESTIONS TO ASK

The following are the questions to ask to improve the use of theory to build models that help understand and optimize product and process designs and develop specifications.

1. Are the reasons for making the most use of available theory well understood or do people think that literature review and thinking through the theory are much work without much return?

2. When designing screening experiments, do people simply attempt to identify the potentially important factors, or do they attempt to postulate the likely effect of each factor, which factors may interact, and propose a mathematical model relating inputs and outputs? For example, do they analyze whether the relationship will be additive, multiplicative, a power function, and so on?

3. Are specifications being developed purely empirically or on the basis of a functional understanding of the relationship between inputs and outputs?

4. Are attempts being made to physically interpret the fitted models to understand the likely action mechanism of the system under consideration?

5. Regression analysis is a commonly used technique for fitting linear and nonlinear equations to data. Are people in your organization sufficiently familiar with regression analysis? Do they understand how to correctly interpret standard error of a parameter, p-values, R^2, standard deviation of residual, and so on to make correct decisions?

Further Reading   The book by Neter et al. (1990) provides excellent detailed information on regression analysis. The accelerated stability testing example is taken from Joglekar (2003).

CHAPTER 11

QUESTIONS AND ANSWERS

The questions and answers in this chapter are intended to provide you an opportunity to test your knowledge of practical statistics, and to help you decide which parts of the book to focus on.

This chapter consists of a test and answers to the test. You can use this book, the book by Joglekar (2003), other books, or computer programs to answer the test. It should be possible to answer most of the questions without using external aids. You could use this test as a pretest, namely, take the test before reading this book. If you answer all questions correctly, you are doing well indeed! If you missed a few (or many) answers, do not be discouraged, this book is for you. Please either read this book from beginning to end or read those portions of the book where you missed the answers. The sequence of questions generally follows the sequence in this book. You could then take the test again.

A green belt should be able to answer most of the questions correctly. A black belt should not only be able to answer the test questions correctly, but also be able to apply the statistical methods to solve industrial problems. A master black belt would be expected to be able to teach these methods, get people enthused about the use of these methods, and answer new questions that may arise.

This multiple-choice test contains 100 questions. Please do not guess. Before you answer a question, make sure that you truly do know the answer and can explain it to others. There is only one correct answer for each question.

Good luck!

Industrial Statistics: Practical Methods and Guidance for Improved Performance   By Anand M. Joglekar
Copyright © 2010 John Wiley & Sons, Inc.

11.1   QUESTIONS

1. Three parts are stacked on each other to produce a stack height. Each part has a normal distribution with mean $\mu = 50$ and standard deviation $\sigma = 2$. What is the standard deviation of stack height?

    a. 6
    b. $2\sqrt{3}$
    c. 12
    d. $3\sqrt{2}$
    e. $4\sqrt{3}$

2. The drug content of three randomly selected tablets turned out to be 48, 50, and 52 mg. What is the sample standard deviation and percent CV of drug content?

    a. 4, 8%
    b. 4, 4%
    c. 2, 1%
    d. 2, 4%
    e. 2, 2%

3. What is the approximate sample size to estimate population standard deviation within 20 percent of the true value with 95 percent confidence?

    a. 10
    b. 20
    c. 50
    d. 100
    e. 200

4. Seven pizzas were randomly selected from a production line and their weights were measured. The sample average was found to be 20 oz and the sample standard deviation was 0.5 oz. What is the approximate interval within which 95 percent of the pizza weights are expected to be with 95 percent confidence?

    a. 20 ± 0.5
    b. 20 ± 1.0
    c. 20 ± 1.5
    d. 20 ± 2.0
    e. 20 ± 2.5

5. For continuous data, the assumption of normality is very important for which of the following tasks?

    a. Calculating variance
    b. Calculating the confidence interval for mean
    c. Conducting a t-test

d. Designing a chart of averages

e. Calculating the tolerance interval

6. The moisture content of a powder was measured at two randomly selected locations inside the drum and was found to be 1.0 percent and 3.0 percent, respectively. What is the estimated standard deviation of the average moisture content inside the drum?

   a. $\sqrt{2}$

   b. 2

   c. 1

   d. $2\sqrt{2}$

   e. $1/\sqrt{2}$

7. Based upon four observations, the sample average was 10 and the sample standard deviation was 0.7. What is the 95 percent confidence interval for the population mean?

   a. 10 ± 1.4

   b. 10 ± 0.7

   c. 10 ± 1.1

   d. 10 ± 0.75

   e. 10 ± 1.5

8. The population standard deviation was expected to be 2.0. How many observations are necessary to estimate the population mean within ± 1.0?

   a. 4

   b. 8

   c. 12

   d. 16

   e. 20

9. A portfolio has an average yearly return of 4 percent with a standard deviation of 2 percent. Average yearly inflation is 3 percent with a standard deviation of 1 percent. What is the standard deviation of real (inflation adjusted) yearly portfolio returns?

   a. 1.0%

   b. 1.6%

   c. 2.2%

   d. 2.8%

   e. 3.4%

10. Large company stocks have an average yearly return of 9 percent with a standard deviation of 12 percent. Small company stocks have an average yearly return of 11 percent with a standard deviation of 16 percent. Under the assumption that there is no correlation between the two asset classes, what

is the standard deviation of the yearly portfolio returns if the portfolio consists of equal proportions of large and small company stocks?

a. 15.5%

b. 14.0%

c. 12.5%

d. 11.5%

e. 10.0%

11. Large company stocks have an average yearly return of 9 percent with a standard deviation of 12 percent. Small company stocks have an average yearly return of 11 percent with a standard deviation of 16 percent. The large and small company stocks are correlated with a correlation coefficient of 0.6. A portfolio has 50 percent large company stocks and 50 percent small company stocks. What is the standard deviation of the yearly portfolio returns?

a. 15.5%

b. 14.0%

c. 12.5%

d. 11.5%

e. 10.0%

12. Raw material viscosity is an important property of interest. There are two raw material suppliers and many viscosity observations are available for material supplied by each supplier. We have to determine if the suppliers *meaningfully* differ in terms of the average viscosity. What is the *best* way to do this?

a. Conduct an *F*-test on the collected data.

b. Conduct a *t*-test on the collected data.

c. Conduct a paired *t*-test on the collected data.

d. Construct a confidence interval for difference of means.

e. None of the above, because viscosity typically does not have a normal distribution.

13. To compare the means of two populations, what is the approximate sample size per population if the practically important difference in means is 0.5, the expected standard deviation is 1.0, the α risk is 5 percent and the β risk is 10 percent?

a. 20

b. 40

c. 60

d. 80

e. 100

14. Which of the following statements is *not likely* to reduce the sample size necessary to compare two population means?

   a. Increase the α risk

   b. Replace attribute data by variable data

   c. Reduce the practical difference to be detected

   d. For attribute data, conduct worst-case testing

   e. Increase the β risk

15. For a certain product there was a one-sided specification for pull force with a lower specification limit of 8.0. Current production had a mean of 10.0 with a standard deviation of 1.0 implying that approximately 2.5 percent of the products did not meet the specification. A designed experiment was being planned to identify the factors to increase the mean pull force. Let Δ denote the practically meaningful effect size to be detected. Which of the following is *not* a consideration in selecting the value of Δ?

   a. The number of factors likely to influence mean

   b. Potential reduction in standard deviation

   c. Impact of out-of-specification product

   d. Whether the experiment is a full factorial design or a fractional factorial design

   e. a, b, and c above

16. Given a practically meaningful difference in means Δ and a confidence interval for difference of means, which of the following statements is *incorrect*?

   a. If the confidence interval includes zero, the difference in means is statistically insignificant.

   b. If the confidence interval includes Δ, more data should be collected before making a decision.

   c. If the confidence interval is completely inside $\pm\Delta$, the difference in means is practically important.

   d. If the confidence interval excludes zero but is within $\pm\Delta$, the t-test would have shown that the difference was statistically significant.

   e. If the confidence interval includes zero and also Δ, the t-test would have shown that the difference was statistically insignificant.

17. A design improvement was made with the intent of reducing standard deviation by 33 percent. For an α risk of 5 percent and a β risk of 10 percent, what is the desired sample size for the old design and the improved design to demonstrate that the variability reduction objective was met?

   a. 10

   b. 30

   c. 50

    d. 70

    e. 100

18. Two formulations were being compared with each other, in terms of the population means, during the research phase of a formulation development project. There were three observations available on one formulation and two on the other. The observed difference in sample averages was 10.0 and the pooled standard deviation of individual values was 2.0. What is the 95 percent confidence interval for the difference in population means?

    a. 10 ± 5.8

    b. 10 ± 4.0

    c. 10 ± 6.0

    d. 10 ± 4.5

    e. 10 ± 5.3

19. We have to replace the current raw material supplier with another, and have to decide how much data to collect on the raw material from the new supplier before making the decision. Which of the following pieces of information is the *least important* in choosing the value of the practically important difference Δ?

    a. Raw material specification limits

    b. Estimate of raw material variability for the current supplier

    c. Implications of an out-of-specification raw material

    d. Whether the specification is one-sided or two-sided

    e. Estimate of raw material mean for the current supplier

20. The sample standard deviation was 0.8 based upon 21 observations on the old process, and was 0.6 based upon 16 observations on the new process. What is the 90 percent confidence interval for the ratio of the two population standard deviations?

    a. 1.2–1.6

    b. 0.5–2.2

    c. 0.2–3.0

    d. 1.1–1.7

    e. 0.9–2.0

21. Which of the following statements is *not true*?

    a. If the sample size is very large, a practically unimportant difference will be detected by a t-test as statistically significant.

    b. If the sample size is very small, a practically important difference will be shown by a t-test as statistically insignificant.

    c. A confidence interval for difference of two means can be interpreted to determine the statistical significance of the difference of means.

d. The necessary sample size can be exactly determined before data collection. Then, if a *t*-test is done with this sample size, statistical significance and practical significance results will match exactly.

e. A confidence interval for difference of means provides greater useful information as compared with a *t*-test.

22. In a fractional factorial design involving four factors A, B, C, and D in eight trials, The effects of factors A, B, and the confounded interaction $AC = BD$ turned out to be important. Which of the following statements is true?

a. The only important factors are A and B.

b. The only important factors are A, B, and C.

c. The only important factors are A, B, and D.

d. All four factors are important.

e. There is not enough information to identify the important factors.

23. The main effects of factors A, B, C, D, and the interactions AB and AC are to be estimated. What is the smallest number of necessary trials?

a. 4

b. 6

c. 7

d. 8

e. 16

24. What is the total number of necessary trials for a screening experiment if the effect size to be detected is 2.0 and the standard deviation of replicates is 1.0?

a. 4

b. 8

c. 16

d. 32

e. 64

25. For a certain response X, the standard deviation turned out to be proportional to the square of the mean. Which is the correct transformation to use before conducting analysis?

a. $Y = 1/X$

b. $Y = 1/\sqrt{X}$

c. $Y = \log X$

d. $Y = \sqrt{X}$

e. $Y = X^2$

26. Following results were obtained in a 2^2 factorial experiment with two factors A and B. Effects larger than 5.0 were statistically and practically important. Which of the following statements is *not true*?

Trial	A	B	Response
1	−1	−1	10
2	1	−1	20
3	−1	1	40
4	1	1	30

a. A is not an important factor.

b. B is an important factor.

c. Factors A and B interact.

d. A is an important factor.

e. Both A and B are important factors.

27. The calculated total number of trials in a designed experiment turned out to be 16. There were three factors to be investigated. Some interactions were expected to be important. While the team thought that the effects will be mostly linear, there was no complete agreement on this question. What is the best experiment to conduct?

a. A 2^3 factorial

b. A 2^3 factorial with a center point

c. A 2^3 factorial replicated once

d. A fractional factorial repeated four times

e. A central composite design

28. In designing an experiment involving frozen pizza, the team identified bake temperature (T), bake time (t), total energy input (E), which is a function of bake time and bake temperature, and which rack the pizza is placed on (R) as potential factors. Which three factors provide the best way to think about this experiment, particularly in selecting factor levels?

a. Temperature, time, and position

b. Temperature, time, and energy

c. Temperature, energy, and position

d. Time, energy, and position

e. None of the above

29. A fractional factorial involving five factors was designed by assigning factor D to the ABC interaction column and factor E to the AB interaction column. There were a total of eight trials. Which of the following statements is *not true*?

a. $AB = CD$

b. $D = CE$

c. $A = BE$

d. $D = ABC$

e. $A = CD$

30. In five different experiments, the following conclusions were reached regarding five different factors. An effect >2 was considered to be practically important. Which of the following statements is *not true*?

Factor	Effect and 95% Confidence Interval
A	0.2 ± 0.1
B	1.0 ± 0.5
C	3.0 ± 0.8
D	3.0 ± 2.0
E	2.0 ± 0.9

a. Factor A is statistically significant.

b. Factor B is practically unimportant.

c. Factor C is practically important.

d. Factor D is practically important.

e. Factor E needs a larger sample size to make a decision.

31. The analysis of a fractional factorial experiment led to the following results. The model in coded units was $y = 20.0 + 2.0A + 1.5AB - 2.0B + C +$ residual. The R^2 was 40 percent. The standard deviation of residual was 1.0. Which of the following statements is *not true*?

a. The measurement variability may be large.

b. If predictions are made using this model, the predicted values of y may be wrong by ± 2.0.

c. An important factor may be missing from the model.

d. The effect of factor A depends upon the level of factor B.

e. The predicted value of y may be wrong by 40 percent.

32. Which of the following considerations is *unimportant* in selecting a response to measure?

a. The response should be easy to measure.

b. The response should be continuous, not discrete.

c. The response should not cause control factor by control factor interactions to occur.

d. The response should not cause control factor by noise factor interactions to occur.

e. The response should have small measurement variability.

33. Which of the following statements is *incorrect* when selecting control factors and their levels when designing a fractional factorial experiment to improve product design?

a. All potentially important factors should be included in the experiment.

b. Factors should be varied over a wide yet realistic range.

c. A center point trial should be included to assess curvature.

d. When multiple factors have the same function, all of them should be included in the experiment.

e. Continuous control factors are preferred.

34. Which of the following effects provide the key to designing robust products?

   a. The main effects of control factors

   b. The interactions between control factors

   c. The main effects of noise factors

   d. The interactions between control and noise factors

   e. The interactions between noise and noise factors

35. A large number of control factor interactions were observed at the product design stage. Which of the following statements is *incorrect*?

   a. There may be problems during scale-up

   b. Wrong response may have been selected causing these interactions to occur unnecessarily

   c. Wrong control factors may have been selected

   d. Wrong noise factors may have been selected

   e. Technical knowledge was not fully utilized in designing the experiment

36. A robustness experiment involved five control factors A, B, C, D, and E. The AB interaction was expected to be large. There were two noise factors and the likely effects of noise factors on product performance were known. The purpose of the experiment was to find levels of the five control factors that will make the product design robust against the noise factors. Which of the following is the best way to design the experiment?

   a. Use a seven factors (five control and two noise) in eight-trial fractional factorial design to evaluate the main effects of the control and noise factors.

   b. Design a separate eight-trial control factor experiment to account for the five main effects and the AB interaction. Design a separate four-trial full factorial in the two noise factors. Evaluate the performance of the eight prototype designs under each of the four noise combinations.

   c. Design a separate eight-trial control factor experiment to account for the five main effects and the AB interaction. Select two combinations of the noise factors to span the noise space. Evaluate the performance of the eight prototype designs under each of the two noise combinations.

   d. Conduct a central composite design in the seven control and noise factors to find all the interactions and quadratic effects.

   e. Conduct a fractional factorial experiment with the seven control and noise factors in 16 trials. This experiment will permit many interactions to be determined.

37. A and B are two components of an electronic circuit. A can have large variability from one product to the next. B is well-controlled. Additionally, the circuit performance is also influenced by ambient temperature T. Which of the effects will *not* be useful in accomplishing robustness?

    a. Quadratic effect of A
    b. Quadratic effect of B
    c. AB interaction
    d. AT interaction
    e. BT interaction

38. The purpose of a designed experiment was to find factors to maximize pull strength y_i. Let μ and σ represent the mean and standard deviation of pull strength and let p be the fraction defective. Which of the following is the correct signal to noise ratio to use?

    a. $-10 \log \left(\dfrac{1}{n} \sum y_i^2 \right)$
    b. $-10 \log \left(\dfrac{1}{n} \sum \dfrac{1}{y_i^2} \right)$
    c. $10 \log \dfrac{\mu^2}{\sigma^2}$
    d. $-10 \log \left(\dfrac{p}{1-p} \right)$
    e. None of the above

39. Data for one of the eight trials of a screening experiment were inadvertently lost. In which of the following situations is this of not much concern?

    a. A replicated factorial in two factors
    b. A factorial in three factors
    c. A four factors in eight-trial fractional factorial
    d. A seven factors in eight-trial saturated fractional factorial
    e. Both a and b above

40. In a robust product design experiment, we need to select control factors, noise factors, and responses. Which is the right sequence in which these selections should be made?

    a. Control factors first, noise factors next, and responses last
    b. Noise factors first, control factors next, and responses last
    c. Responses first, control factors next, and noise factors last
    d. Responses first, noise factors next, and control factors last
    e. The sequence does not matter

41. Noise factors may belong to which of the following categories?

    a. External noise due to factors external to the system such as ambient temperature or humidity

    b. Internal noise or product-to-product variation caused by manufacturing

    c. Product degradation over time

    d. Replication variability caused by internal or external factors

    e. All of the above

42. Which of the following statements is correct?

    a. Traditionally, specifications are interpreted to mean that a product inside the specification is good and product outside the specification is bad.

    b. It is almost impossible to distinguish between a product that is just inside the specification and another that is just outside the specification.

    c. As the product deviates from target, product quality deteriorates even inside the specification.

    d. As the product deviates from target, the bad effects on the customers increase more than proportionately.

    e. All of the above statements are true.

43. Products made from 13 raw material lots were found to be satisfactory. The moisture content in the 13 raw material lots was measured and the average was found to be 5 percent and the standard deviation was 0.5 percent. What is the empirical two-sided specification for moisture that includes 99 percent of the population with 95 percent confidence?

    a. 5 ± 0.5

    b. 5 ± 1.0

    c. 5 ± 2.0

    d. 5 ± 1.5

    e. 5 ± 3.0

44. Three parts were stacked on each other to produce a stack height. The specification for each part is 2 ± 0.1. What is the worst-case specification for stack height?

    a. 4 ± 0.2

    b. $6 \pm \sqrt{0.3}$

    c. $4 \pm \sqrt{0.2}$

    d. 6 ± 0.3

    e. 6 ± 0.6

45. Two similar parts were stacked on top of each other to produce a stack height. The statistical specification for stack height had a target of 10.0 and a standard deviation of 1.0 denoted by $10 < 1$. What is the statistical specification for the part?

    a. $5 < 1.5$

    b. $5 < 0.5$

c. $5 < 0.7$

d. $5 < 1.4$

e. $5 < 1.0$

46. Two similar parts were stacked on top of each other to produce a stack height. The worst-case specification for stack height is 10.0 ± 3.0. What is the worst-case specification for the part?

a. 5 ± 1.5

b. 5 ± 0.5

c. 5 ± 0.7

d. 5 ± 1.4

e. 5 ± 1.0

47. A deviation of 10 volts from target causes the voltage converter to malfunction leading to a customer loss of $100. Each converter is inspected before shipment and the manufacturer can adjust the voltage at a cost of $4. What should be the producer's voltage specification around the target?

a. ± 1

b. ± 2

c. ± 3

d. ± 4

e. ± 5

48. A company manufactures machines to analyze blood chemistry. A hospital may purchase one or two of these machines, which they will use for several years to come. Which approach should be used to set specifications on important machine performance characteristics?

a. Empirical approach

b. Functional—worst-case approach

c. Functional—statistical approach

d. Functional—unified approach

e. Any of the above approaches could be used

49. The specification for an important characteristic is ± 4 around the target. The product is currently in development and the job of developing the product and the job of controlling the manufacturing process are judged to be equally difficult. The company wants to ensure that the P_p index in manufacturing will be at least 1.33. What should be the standard deviation targets for R&D and for manufacturing?

a. 0.5 and 1.0

b. 1.0 and 0.5

c. 2.0 and 2.0

d. 0.7 and 0.7

e. 0.55 and 0.55

50. Regarding setting specifications, which of the following statements is *not true*?

    a. Worst-case specifications on inputs lead to a wider specification on output compared with the statistical approach.

    b. Statistical specification on output leads to wider specifications on inputs compared with the worst-case approach.

    c. Statistical specifications on inputs lead to tighter specification on output compared with the worst-case approach.

    d. Worst-case specification on output leads to tighter specifications on inputs compared with the statistical approach.

    e. Statistical specifications on inputs lead to wider specification on output compared with the worst-case approach.

51. Two input factors A and B are related to the output Y as a multiplicative function $Y=AB$. The specification for Y is 100 ± 10. The targets for A and B are 20 and 5. Which of the following may be a reasonable specification for A and B, respectively?

    a. 20 ± 2.0 and 5 ± 0.25

    b. 20 ± 0.8 and 5 ± 0.8

    c. 20 ± 0.4 and 5 ± 0.4

    d. 20 ± 0.2 and 5 ± 0.2

    e. 20 ± 0.1 and 5 ± 0.1

52. Acceptable quality level (AQL) refers to which of the following?

    a. The number of samples pulled to execute the acceptance sampling plan.

    b. Acceptable risk of rejecting a good lot.

    c. Acceptable risk of accepting a bad lot.

    d. Largest percent defective acceptable as a process average.

    e. The acceptance number of a sampling plan.

53. Which of the following is a true statement?

    a. An accepted lot is a good lot.

    b. Acceptance sampling is counterproductive if the manufacturing process is stable.

    c. A rejected lot is a bad lot.

    d. Accept-on-zero plan guarantees that all accepted lots have zero defectives.

    e. Producers should decide AQL and consumers should decide RQL.

54. If zero defectives are found in a sample of size 50, we can say with 95 percent confidence that the true percent defective is less than or equal to which of the following?

a. 3%

b. 1%

c. 0%

d. 10%

e. 6%

55. The AQL and RQL to be used in manufacturing are 1 and 5 percent, respectively. Which of the following statements is true?

a. The AQL in validation should be 1 percent.

b. The RQL in validation should be 5 percent.

c. The RQL in validation should be 1 percent.

d. The AQL in validation should be 5 percent.

e. None of the above.

56. An attribute acceptance sampling plan has a sample size of 10 and an acceptance number of 0. What is the probability that a lot containing 1 percent defectives will be accepted by the sampling plan?

a. 80%

b. 85%

c. 90%

d. 95%

e. 100%

57. Which of the following statements is *false*?

a. A critical characteristic should have a small AQL.

b. A critical characteristic should have a small RQL.

c. If several lots in a sequence fail, the sampling plan should be changed to reduce both AQL and RQL.

d. If CQL, the fraction defective routinely produced by the manufacturing process is less than AQL, many lots will fail.

e. AQL and RQL can be selected based upon industry practice.

58. If a sampling plan has been proposed by the customer, the first action that the supplier should take is which of the following?

a. Accept the plan as proposed.

b. Propose an alternate plan.

c. Determine the AQL for the proposed plan.

d. Determine the RQL for the proposed plan.

e. Determine CQL, the fraction defective routinely produced by the manufacturing process, and assess the corresponding risk of lot rejection.

59. A certain sampling plan has the following operating characteristic curve. Which of the following statements is false?

Operating characteristic (OC) curve
n=100, c=2

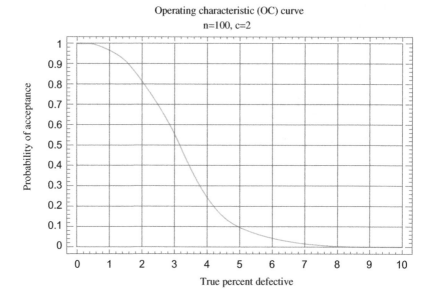

a. The AQL is approximately 0.6 percent.

b. The RQL is approximately 5 percent.

c. A lot with 3 percent defectives will be rejected 60 percent of the time.

d. A lot with 6 percent defectives will be accepted 5 percent of the time.

e. As long as the current quality level (CQL) (the fraction defective currently produced by the process) exceeds AQL, the probability of lot failure is small.

60. One company used the following approach to accept or reject a lot. If there were no defectives in a sample of size 10, the lot was accepted. For the rejected lot, a second sample of size 10 was taken and if no defectives were found, the lot was accepted. Otherwise, the lot was rejected. What is the probability of accepting a lot with 5 percent defectives?

a. 60%

b. 72%

c. 78%

d. 84%

e. 89%

61. A selection is to be made between an attribute single sample plan, a variable single sample plan, and an attribute double sample plan. All three are matched plans, namely, they have the same operating characteristic curve. Which of the following statements is *not* true?

a. A double sampling plan may be difficult to administer.

b. An attribute double sampling plan will have smaller expected sample size than the attribute single sampling plan.

c. A variable single sampling plan will have a smaller sample size than an attribute single sampling plan.

d. A variable sampling plan will provide greater protection than an attribute sampling plan.

e. The cost of taking variable measurements could be considerably higher than the cost of attribute measurements.

62. Which of the following concepts is *not* used in determining the control limits for a chart of averages with a subgroup size of two?

a. Centerline is often taken to be the grand average.

b. Standard deviation of the average reduces inversely proportional to the square root of the subgroup size.

c. The data are normally distributed.

d. Between-subgroup variance is used to calculate control limits.

e. False alarm probability is 3 in a 1000.

63. Which is the *least* important factor in determining the sampling interval for a control chart?

a. Expected frequency of process shifts

b. Cost of bad product

c. Capability index

d. Process mean

e. Cost of sampling and testing

64. Which is the *least* important factor in determining the subgroup size for a control chart?

a. The risk of not catching a process shift

b. The risk of false alarm

c. Cost of sampling and testing

d. Cost of bad product

e. Capability index

65. For a chart of averages with a subgroup size of 4, the upper control limit UCL $= 12$ and the lower control limit LCL $= 8$. What will be the control limits if the subgroup size is changed to one?

a. UCL $= 11$ and LCL $= 9$

b. UCL $= 12$ and LCL $= 8$

c. UCL $= 13$ and LCL $= 7$

d. UCL $= 14$ and LCL $= 6$

e. UCL $= 15$ and LCL $= 5$

66. When should a control chart be redesigned?
  a. When it produces an out-of-control signal
  b. Whenever new data become available, in fact the control limits should be recalculated with each additional data point
  c. When a deliberate process change has been made and the control chart indicates that the process has changed
  d. Both a and c above
  e. Both a and b above

67. What does a β risk mean for a control chart?
  a. The risk of not detecting a shift in the mean
  b. The risk of producing bad product
  c. The risk of false alarm
  d. The risk of misclassifying a product
  e. The risk of concluding that a process is out-of-control when it is not

68. Which of the following is *not* an out-of-control rule?
  a. One point outside the control limits
  b. Four points in a row below the centerline
  c. Two out of three successive points outside the two standard deviation limits
  d. Six points in a row going up
  e. 14 points alternating up and down

69. Given $C_p = 1.5$, $C_{pk} = 1.5$, and $P_{pk} = 0.8$, which corrective action should be taken?
  a. Stabilize the process
  b. Center the process
  c. Fundamentally change the process
  d. Centering and stabilizing the process are equally important
  e. No action is necessary since $C_p = C_{pk}$

70. Which of the following assumptions is necessary for the C_{pk} index, but not for the P_{pk} index?
  a. The distribution is normal.
  b. The process is centered.
  c. The specification is known.
  d. The process is stable.
  e. The specification is two-sided.

71. What will you conclude if the computed P_{pk} is negative?
  a. The computation is wrong.
  b. Process average is outside the specification.

    c. Process average is less than zero.

    d. The total standard deviation is much larger than the short-term standard deviation.

    e. One specification limit is negative.

72. If the process is improved by centering it and reducing its variability, what will happen?

    a. P_p will increase

    b. P_{pk} will increase

    c. P_p and P_{pk} will be equal

    d. a and b above

    e. All of the above

73. If the lower specification limit is 2, the upper specification limit is 10, process mean is 8, and within-subgroup standard deviation is 2, what is the C_{pk} index?

    a. 0.33

    b. 0.50

    c. 1.00

    d. 1.33

    e. None of the above

74. Which of the following statements is *incorrect*?

    a. A stable process is a predictable process.

    b. A stable process is a capable process.

    c. A process may have $P_p = 1.3$ and $P_{pk} = 0.8$.

    d. Both a and b above.

    e. Both a and c above.

75. Which of the following is *not* the purpose of variance components analysis?

    a. Determine the variance between raw material batch means

    b. Assess within-subgroup variance

    c. Assess between-subgroup variance

    d. Determine repeatability and reproducibility of a measurement system

    e. Assess if there is a linear time trend to the data

76. Two observations each were taken on two batches of raw material. For the first batch, the observations were 8 and 9 and for the second batch, the observations were 10 and 11. What are the within- and between-batch variance estimates?

    a. Within-batch variance $= 0.50$ and between-batch variance $= 1.75$

    b. Within-batch variance $= 0.50$ and between-batch variance $= 1.50$

    c. Within-batch variance $= 0.50$ and between-batch variance $= 2.00$

    d. Within-batch variance $= 0.25$ and between-batch variance $= 1.00$

    e. Within-batch variance $= 0.25$ and between-batch variance $= 0.75$

77. For a chart of averages, the variance within subgroup was found to be 1.0 and the variance between subgroups was found to be 0.2. What action may significantly reduce variability?

a. Eliminate shift-to-shift differences

b. Work with the supplier to reduce raw material batch-to-batch variability

c. Institute humidity control to remove the seasonal humidity variation

d. Remove the time trend from one subgroup to the next by using a control chart

e. Evaluate measurement system variability

78. Average quadratic loss due to a perfectly centered process is $2 million per year. If the standard deviation of the process is reduced by a factor of 2, what will be the new average loss per year?

a. $1 million

b. $2 million

c. $0.75 million

d. $0.25 million

e. None of the above

79. Ambient humidity influences process output. Experiments were purposefully conducted on a very low humidity day and a very high humidity day, and the contribution of humidity to the variance of output was determined assuming the two days to be a random sample from the population of humidity. Which of the following statements is true?

a. Variance contribution due to humidity is correctly estimated because the average humidity in the experiment is close to true average humidity throughout the year

b. Variance contribution due to humidity is overestimated

c. Variance contribution due to humidity is underestimated

d. Variance contribution due to humidity is correctly estimated because of the central limit theorem

e. It is not possible to say if the variance contribution due to humidity is correctly estimated, overestimated, or underestimated

80. A variance component study resulted in the following information. Process mean $= 8$, raw material batch-to-batch variance $= 2$, manufacturing variance 4, and test variance $= 1$. The specification was 10 ± 4. Which *single action* will improve the P_{pk} index by the largest amount?

a. Center the process

b. Eliminate raw material batch-to-batch variance

c. Eliminate manufacturing variance

d. Eliminate test variance

e. Eliminate both raw material and test variance

81. A company used three similar machines to produce products. These were the only machines of this type that the company had. To determine the differences between the three machines, 100 products made by each machine were evaluated. Which of the following statements is true?

a. Machines is a fixed factor

b. Products are nested inside machines

c. Machines are nested inside products

d. Machines and products are crossed

e. Both a and b above

82. How will you interpret the following variance components table based upon data collected for a chart of averages?

Source of Variation	Degrees of Freedom	Variance
Between subgroups	25	0.2
Within subgroup	100	1.0

a. The first action to take is to continue to control chart the process and remove special causes as they are found

b. The total variance is too small

c. The common cause variance is 1.0

d. The common cause variance is 0.2

e. The degrees of freedom are insufficient

83. A pizza chain had four restaurants in town. Each of six randomly selected people visited each of the four restaurants. Each person ordered two cheese pizzas and evaluated their quality. What is the structure of this study?

a. People are nested inside restaurants and pizzas are nested inside people

b. People and restaurants are crossed and pizzas are nested

c. All three factors are crossed

d. People and pizzas are crossed, but both are nested inside restaurants

e. None of the above

84. If W represents the half-width of a two-sided specification, the total standard deviation of the measurement system should be less than which of the following for the measurement system to be acceptable for specification-related measurements?

a. W

b. W/2

c. W/5

d. W/10

e. W/20

85. If σ_t represents the total standard deviation of the process, the measurement standard deviation should be less than which of the following for the measurement system to be satisfactory for product and process improvement purposes?

   a. σ_t

   b. $\sigma_t/3$

   c. $\sigma_t/5$

   d. $\sigma_t/10$

   e. None of the above

86. Why are the acceptance criteria for the measurement system standard deviation wider for destructive tests compared with nondestructive tests?

   a. Destructive tests are more expensive

   b. For destructive tests, the sample sizes have to be smaller

   c. For destructive tests, the measurement standard deviation includes the short-term product-to-product standard deviation

   d. Nondestructive tests permit multiple measurements of the same product

   e. None of the above

87. Three radically different approaches were proposed to measure drug content x. The outputs of the three different measurement approaches are denoted by y_1, y_2, and y_3, respectively. Each approach was calibrated using drug content standards resulting in three different slopes (d_y/d_x) for the linear calibration curves and three different estimates of the measurement standard deviation σ_y. Which of the three proposed measurement approaches is the best?

   a. The one with the largest slope

   b. The one with the smallest standard deviation

   c. The one with the largest ratio of slope/standard deviation

   d. The one with the largest ratio of standard deviation/slope

   e. None of the above

88. Four products were measured in a measurement system study. Two operators were randomly selected from a large pool of operators to measure the first product, another two were randomly selected to measure the second product, and so on. Each operator made duplicate measurements on each product measured by that operator. What is the structure of this measurement system study?

   a. Duplicates are nested inside operators and operators are nested inside products

   b. Duplicates and products are crossed and operators are nested

   c. Products and operators are crossed and duplicates are nested

   d. Duplicates and operators are crossed and products are nested

   e. All three factors are crossed

89. In a Gage R&R study, the interaction between the operators and products turned out to be the largest variance component. What is the right corrective action?

    a. Reduce the inherent variability of the measurement system

    b. Control chart the measurements to ensure that there is no time trend

    c. Train the operators to measure alike

    d. Repeat the measurement study with products that are more alike

    e. None of the above

90. A large number of factors were evaluated in a designed experiment conducted to validate the measurement system. The measurement standard deviation was estimated to be 0.1. One of the factors, factor A, also turned out to have a statistically significant effect. What can we conclude from this information?

    a. The measurement system is acceptable for product and process improvement purposes.

    b. The measurement system is acceptable for specification-related measurements.

    c. The measurement system is not robust.

    d. The specification for factor A must be constrained.

    e. None of the above.

91. Suppliers A, B, and C have the same $P_{pk} = 1.2$. The measurement standard deviation is large for supplier A, medium for supplier B, and small for supplier C. Who will you buy the product from?

    a. Supplier A

    b. Supplier B

    c. Supplier C

    d. All are equally good because they have the same P_{pk}

    e. None, because the P_{pk} is less than 2.0 and does not meet the 6-sigma target

92. A Gage R&R study resulted in the conclusions summarized in the variance components table below. Which of the following statements is true?

Source of Variability	Variance Due to Each Source as % of Total Variance
Product to product	90
Operator to operator	5
Operator by product interaction	1
Instrument	4
Total	100

    a. The measurement system is suitable for specification-related purposes

    b. Operators need training because the operator-to-operator variability exceeds the instrument variability

    c. The measurement system is suitable for product and process improvement purposes

    d. The product-to-product variability is too large

    e. None of the above

93. When should a measurement system be adjusted?

    a. As often as economically feasible

    b. Everyday at the beginning of a shift

    c. Whenever the operator changes

    d. When a control chart of measurements of a standard over time signals out of control

    e. At fixed intervals as a preventive measure

94. Let R denote the syringe radius, L the stroke length, and B the volume of back flow. The volume of fluid injected per stroke is equal to which of the following?

    a. $\pi R^2 L$

    b. $\pi R^2 L - B$

    c. $2\pi RL - B$

    d. $B - 2\pi RL$

    e. None of the above

95. The relationship between output y and inputs A and B is given by $y = A\sqrt{B}$. A has a target of 10 and a standard deviation of 1.0. B has a target of 4 and a standard deviation of 1.0. What is the CV for y?

    a. 12%

    b. 16%

    c. 20%

    d. 22%

    e. 25%

96. Regression analysis was used to obtain the following equation relating output y to input factors A and B.

$$y = 100 - 5A + B + \text{residual}$$

Both coefficients in the equation were statistically significant at 95 percent confidence and the R^2 was 25 percent. Which of the following statements is *not* true?

    a. A influences the output

    b. B influences the output

    c. Some other factor may be influencing the output

    d. Measurement variability may be large

e. Measurement variability must be small for the effects of A and B to turn out to be significant

97. Which of the following statements is *not* true? For a linear regression expressing output y as a linear function of input x, R^2 will be large if

a. The range of x is large.

b. The measurement error is small.

c. The range of x is small.

d. No other factor influences the output.

e. None of the above.

98. The expected relationship between output y and inputs A and B is $y = aA^b B^c$. A literature review would have identified this relationship, but the review was not done. Instead, a designed experiment is planned to determine the relationship between inputs and output. Which of the following statements is true?

a. The designed experiment could lead to the wrong model.

b. An unnecessarily large designed experiment may be planned.

c. The fitted model may result in large prediction errors.

d. The fitted model may result in inadequate scientific understanding of the subject.

e. All of the above.

99. Based upon 1000 observations over time, the collected moisture data turned out to have a significant positive autocorrelation. Which of the following statements is *incorrect* for autocorrelated data?

a. Future values of moisture cannot be predicted with confidence

b. The out-of-control rule *eight points in a row above the center line* will not apply

c. The mean cannot be well estimated

d. The out-of-control rule *six points going up makes a trend* will not apply

e. The data will exhibit cyclic behavior

100. The following information was obtained for a linear relationship between an input and an output: prediction error standard deviation $= 1.5$, $R^2 = 88$ percent, confidence interval for the intercept $= 0.2 \pm 0.5$, and confidence interval for slope $= 1.2 \pm 0.3$. Which of the following statements is *incorrect*?

a. The fitted straight line may go through the origin

b. The slope is statistically significant

c. This input factor captures 88 percent of the variance of output

d. If this equation is used to predict the output, the predicted values could be wrong by ± 3.0

e. The fitted equation appears to be a very useful equation

11.2 ANSWERS

1. **(b)** The mathematics of standard deviation is complicated but the mathematics of variance is simple. Variances add. The variance of each part is 4. Therefore, the variance of three parts stacked on top of each other is $(4+4+4) = 12 = 2\sqrt{3}$.

2. **(d)** The sample mean and variance calculated from the three data values 48, 50, and 52 are 50 and 4, respectively. Therefore, the standard deviation is 2 and the percent CV is $(2/50)100 = 4$ percent.

3. **(c)** The sample size to estimate σ within δ percent of the true value with 95 percent confidence is approximately given by

    $$n = 2\left(\frac{100}{\delta}\right)^2$$

    So for $\delta = 20$, the necessary sample size is 50.

4. **(d)** This is not the case of constructing a confidence interval for mean. We have to construct a tolerance interval for individual values because the question relates to the weights of individual pizzas. This tolerance interval is given by sample average $\pm k$ sample standard deviation. The value of k depends upon percent confidence (95 percent), percent of the population enclosed in the interval (95 percent), and the sample size (7), and is 4.007 from the table of k values in the Appendix. For a sample average of 20 and a sample standard deviation of 0.5, the tolerance interval is 20 ± 2.0.

5. **(e)** The calculation of variance does not make any assumptions regarding the distribution. Due to central limit theorem, the assumption of normality is less important when dealing with sample averages involving sample sizes greater than six. The assumption of normality becomes very important when inferences are to be drawn regarding the tails of the distribution. Therefore, the assumption of normality is not crucial for computing the confidence interval for mean, for conducting a t-test that concerns differences in means, and for designing a chart of averages. Lack of normality will significantly influence the calculated tolerance interval because the calculation has to do with the tails of the distribution of individual values.

6. **(c)** The variance of *individual values*, calculated from the two observations 1 and 3 percent is $\sqrt{2}\%$. The standard deviation of sample averages reduces by a factor of the square root of sample size and is equal to 1 percent.

7. **(c)** The 95 percent confidence interval for mean is given by $\bar{x} \pm ts/\sqrt{n}$. From the Appendix, the t value, corresponding to three degrees of freedom and a tail area probability of 0.025 is 3.182. Hence, the answer is 10 ± 1.1.

8. **(d)** The 95 percent confidence interval for mean is $\mu \pm 2\sigma/\sqrt{n}$. Hence, we expect $2\sigma/\sqrt{n} = 1.0$. Given that the population standard deviation is 2.0, $n = 16$.

9. **(c)** Let X represent the yearly portfolio returns with an average of 4 percent and a standard deviation of 2 percent. Let Y represent yearly inflation with an average of 3 percent and a standard deviation of 1 percent. Then, real returns correspond to $(X - Y)$. The average yearly real return is 1 percent. The variance of yearly real returns is $\text{Variance}(X - Y) = \text{Variance}(X) + \text{Variance}(Y) = 5$ (percent)^2. Thus, the standard deviation of yearly real returns is 2.2 percent.

10. **(e)** Large company stocks (X) have an average yearly return of 9 percent with a standard deviation of 12 percent. Small company stocks (Y) have an average yearly return of 11 percent with a standard deviation of 16 percent. Let a be the fraction of the portfolio in large company stocks. Under the assumption that the large and small company stocks are uncorrelated,

$$\text{Variance}(aX + (1-a)Y) = a^2\text{Variance } X + (1-a)^2\text{Variance } Y$$

For our example, $a = 0.5$, and substituting in the above equation, the portfolio variance is 100 or the standard deviation is 10 percent. Note that the standard deviation of the portfolio is smaller than that of the individual asset classes.

11. **(c)** This example is similar to the one above except that the large and small company returns are correlated with a correlation coefficient $\rho = 0.6$. The variance equation changes to

$$\text{Variance}(aX + (1-a)Y) = a^2\text{Variance } X + (1-a)^2$$
Variance $Y + 2a(1-a)\rho(\text{standard deviation of } X)(\text{standard deviation of } Y)$

Upon substitution, the portfolio standard deviation is 12.5 percent. Note that due to the positive correlation between the two asset classes, the portfolio standard deviation has increased compared with what it was in the previous example.

12. **(d)** The two suppliers may differ from each other in terms of the average viscosity and/or the standard deviation of viscosity. This specific question is regarding whether the two suppliers differ in terms of the average viscosity. F-test is used to compare two variances. t-Test is used to compare two means. A paired t-test is used to compare two means when the observations can be meaningfully paired. This is not the case here. The assumption of normality is not crucial here since we are comparing means. The usual approach to solve these problems is to first conduct an F-test to assess if the two variances are statistically the same or not; and then to conduct the appropriate t-test either assuming the variances to be equal or unequal. As explained in Chapter 2 of this

book, this usual approach has many drawbacks. In particular, it does not provide sufficient useful information to judge if the difference in means is practically important or not. The right approach is to construct a confidence interval for difference of means.

13. **(d)** For moderate risks of making wrong decisions, the sample size per population is approximately given by

$$n = \frac{20}{d^2}$$

where d is the ratio of practically meaningful difference to be detected to the standard deviation $= 0.5/1.0 = 0.5$. Hence, $n = 80$.

14. **(c)** If we are willing to take larger risks of making wrong decisions, we will not need as much data and the necessary sample size will reduce. If attribute data are replaced by variable data, the sample size will reduce because variable data have greater information. For attribute data, conducting worst-case testing will magnify the differences between the two populations being compared, and the sample size will reduce. If the practical difference to be detected becomes smaller, the sample size will increase because more data will be necessary to detect small differences. From the approximate sample size formula given above, if the difference to be detected becomes half, the sample size will increase by a factor of four.

15. **(d)** The current mean is two standard deviations away from the lower specification limit and hence, the percent out-of-specification product is 2.5 percent. To reduce the nonconforming product the mean has to be increased, the standard deviation needs to be reduced, or both. Suppose the objective is to increase the mean so that it is three standard deviations away from the lower specification limit. Since the standard deviation is 1.0, the mean needs to be increased by 1.0. If there is only one factor that may increase the mean then the practically meaningful effect size Δ will be close to 1.0. On the other hand, if there are several factors that may increase the mean, each may contribute a small portion of the desired change and Δ will be smaller because we will be interested in detecting smaller effects that may add up to the desired change. If the standard deviation could be reduced, the mean may not have to be increased by 1.0 and Δ will be smaller. If the impact of out-of-specification product is large, the mean may have to be more than three standard deviations away from the lower specification limit and the value of Δ will be >1.0. The total number of necessary trials is governed by the selected Δ and the standard deviation. Whether the experiment is a full factorial or a fractional factorial has no bearing on the selection of Δ.

16. **(c)** The essential arguments are the following: if the confidence interval for difference of means includes zero, the difference in the two means is

statistically insignificant. If the confidence interval is contained within $\pm\Delta$, then the difference between the two means is practically unimportant. If the confidence interval includes Δ, additional data need to be collected before making a decision. If the confidence interval excludes zero, but is completely within $\pm\Delta$, then the difference in means is statistically significant, but practically unimportant. A decision purely based on the t-test would have been to say that the two means are different. If the confidence interval includes zero and also Δ, the t-test would have shown that the difference was statistically insignificant, however, the real conclusion is that more data are necessary to decide if the difference is practically important or not.

17. **(c)** A reduction of 33 percent in standard deviation means that the ratio of two standard deviations is 1.5. There is no simple explicit formula to compute the sample size. The necessary sample size is obtained from the table in Chapter 2 and the nomogram given in Joglekar (2003). The sample size is 50. A large sample size is necessary to detect small changes in standard deviation. Some sample sizes are worth remembering: to detect a $2:1$ change in standard deviation requires 20 observations per population, a $1.5:1.0$ change requires 50 observations per population, and a $1.3:1.0$ change requires 100 observations per population.

18. **(a)** The 95 percent confidence interval for the difference of means is given by

$$(\bar{x}_1 - \bar{x}_2) \pm t_{\alpha/2, n_1 + n_2 - 2} s_{\text{pooled}} \sqrt{\frac{1}{n_1} + \frac{1}{n_2}}$$

Substituting the values,

$$10 \pm (3.182)(2.0)\sqrt{\frac{1}{3} + \frac{1}{2}} = 10 \pm 5.8$$

The interpretation is that we are 95 percent sure that the two population means may differ by as little as 4.2 or by as much as 15.8.

19. **(d)** When one supplier is replaced by another, there is bound to be some difference between the two population means, however small the difference may be. We have to demonstrate that the difference in the population means is smaller than some practically important difference Δ. The selection of Δ is based upon several considerations. The raw material specification is important to know because we do not want Δ to be so large that a large proportion of product will be outside the specification. An estimate of raw material variability is important in calculating the probability of producing the non-conforming product for various values of Δ. The implications of the non-conforming product need to be understood to judge how much nonconforming product may be acceptable. An estimate of raw material mean for the current supplier is important to know because Δ is being judged in relation to this mean. What is unimportant is whether the specification is one-sided or two-sided.

20. (e) The $100\,(1-\alpha)$ percent confidence interval for σ_1^2/σ_2^2 is

$$\frac{\left(s_1^2/s_2^2\right)}{F_{\alpha/2,n_2-1,n_1-1}} \text{ to } \left(s_1^2/s_2^2\right)F_{\alpha/2,n_1-1,n_2-1}$$

Confidence interval for σ_1/σ_2 is computed by taking square roots. For the current example, $n_1 = 21$, $n_2 = 16$, and for $\alpha = 10$ percent, the 90 percent confidence interval for σ_1^2/σ_2^2 is

$$\frac{(0.8^2/0.6^2)}{2.2} \text{ to } (0.8^2/0.6^2) \times 2.33 = 0.8 \text{ to } 4.14$$

The 90 percent confidence interval for σ_1/σ_2 is 0.89–2.03, obtained by taking square roots. Since the confidence interval for σ_1/σ_2 includes 1, there is no conclusive evidence that the new design has reduced variability. Also, the confidence interval is very wide so that, with the number of observations gathered, a rather large change in standard deviation will go undetected. More data are necessary to make a reasoned decision.

21. (d) The results of a t-test can be gamed! A very large sample size will always lead to the conclusion that the two means are statistically different. A very small sample size will usually conclude that the two means are statistically the same. A confidence interval for difference of means avoids these difficulties and also provides greater information than a t-test regarding the potential magnitude of the difference in means. It is incorrect to think that sample size can be exactly determined before data collection. This is so because the determination of sample size is based upon our estimate of the true difference in means and the true standard deviation. Both these quantities are unknown. If they were known precisely, no data collection will be necessary.

22. (e) An analysis of the fractional factorial design has shown that the main effects of factors A and B; and the confounded interaction AC = BD are important. Based upon this information alone, it is not possible to say whether it is the AC interaction or it is the BD interaction that is the important interaction. A factor is considered to be an important factor if it produces a large main effect or is involved in a large interaction. Hence, in this case, there is not enough information to identify key factors.

23. **(c)** The smallest number of trials is always one more than the number of effects to be estimated. For example, to estimate the main effect of factor A, we need two trials. To estimate the main effects of factors A and B, we need three trials. To also estimate the interaction between A and B, we need four trials, and so on. In this example, there are four main effects and two interaction effects to be estimated. The minimum number of trials is seven. This seven-trial experiment can be constructed, but for various reasons, it may be better to use the eight-trial fractional factorial design.

24. **(c)** The total number of trials in a designed experiment is given by

$$n = \frac{64}{d^2}$$

where $d = \Delta/\sigma$. The smallest practically important difference to be detected is Δ and σ is the standard deviation of replicates. In this example, $\Delta = 2$ and $\sigma = 1$. Hence, $d = 2$ and $n = 16$.

25. **(a)** An important assumption in statistical data analysis is that the mean and standard deviation are uncorrelated. If this is not the case, the data need to be transformed before analysis such that for the transformed data, mean and standard deviation are uncorrelated. The necessary transformations were worked out by Box and Cox, and are known as the Box–Cox transformations. If standard deviation is proportional to the square of the mean, then the transformation to use is the reciprocal transformation. This means that we should analyze the reciprocals of the collected data and after the analysis is done make the inverse transformation to present the results in a familiar metric.

26. **(a)** The main and interaction effects, calculated from the designed experiment are the following: main effect of $A = 0$, main effect of $B = 20$, and interaction $AB = -10$. Even though the main effect of A is zero, it is wrong to conclude that factor A is not important because it is involved with an interaction that is practically and statistically important.

27. **(e)** A 2^3 factorial will determine all main and interaction effects, but not quadratic effects. Since the total number of necessary trials is 16, the factorial will have to be repeated. If a center point is added to the replicated factorial, we will get an indication of the presence of curvature, but not the factor(s) causing it. A fractional factorial replicated four times is a particularly bad choice in this case because the interactions will be confounded and the curvature effects cannot be determined. Since 16 trials are necessary to properly estimate effects, and quadratic effects may be present, a central composite design with 16 trials including a replicated center point is a good choice.

28. **(c)** In the pizza baking experiment described in this question, what matters most is the energy. The right amount of energy can be obtained with a high temperature and low time combination or with a low temperature and high time combination and an infinite number of combinations in between. This means that time and temperature will interact, requiring a different range of time as a function of temperature. It is also possible that at a secondary level the rate of input of energy matters and this is a function of temperature. Hence, the right way to think about this experiment is in terms of temperature, energy, and position, particularly to select factor levels to eliminate the time–temperature interaction in the experiment.

29. **(e)** The experiment was generated with $D = ABC$ and $E = AB$. This means that the identity is $I = ABCD = ABE = CDE$. The confounding structure does include $AB = CD$, $D = CE$, $A = BE$, and $D = ABC$. $A = CD$ is incorrect.

30. **(d)** Factor A is statistically significant because the confidence interval does not include zero. Factor B is practically unimportant because the effect of B is between 0.5 and 1.5 and is not greater than the practically important effect size of 2. Factor C is practically important because its effect is greater than 2. The effect of factor D is between 1 and 5. Factor D may be practically important or may not be. More data are needed to establish the truth. The case with factor E is similar.

31. **(e)** The fact that $R^2 = 40$ percent means that 60 percent of the variance of response is not explained by the factors in the regression equation. Something else is causing this variability. One reason could be measurement variability. Another reason could be that an important factor is missing from the equation. Based upon this equation, the 95 percent prediction interval is approximately ± 2. The effect of factor A depends upon factor B because there is an AB interaction. The statement that the predicted value of response could be wrong by 40 percent is incorrect, it could be wrong by ± 2.

32. **(d)** Ease of measurement, measurements on a continuous scale, lack of induced control factor interactions, and small measurement variability are good characteristics for a response to have. The statement that a response should not cause control by noise factor interactions is wrong. The control by noise factor interactions are a pre-requisite for robust product and process design.

33. **(d)** All potentially important factors should be included in an experiment because, otherwise, a truly important factor may be missed. The factors should be varied over a wide yet realistic range to make it easier to determine effects. A center point should be included to assess the presence of quadratic effects. Continuous control factors are preferred compared with discrete factors because they allow two-level experimentation, thus, reducing the size of the experiment. When multiple control factors have the same function, they will interact with each other, necessitating a very large experiment and difficult interpretation of results. Only one such factor should be included in the experiment, levels of one factor should be made a function of another, or some alternate ways of formulating the problem should be found.

34. **(d)** The keys to robustness are interactions between control and noise factors. If these interactions exist, it means that the effect of noise factors depends upon the levels of control factors. This makes it possible to reduce the effects of noise factors by the judicious choice of control factor levels, thus making product and process designs robust against noise.

35. **(d)** If there are many control factor interactions at the product design stage, then these factors may also interact with scale-up factors and the optimum design at the bench scale will no longer be optimum when scaled up. These interactions can also occur if a wrong response is selected or wrong control factors are selected (e.g., factors that have an identical function will interact). This means that the right technical knowledge either did not exist or was not fully utilized. Selection of wrong noise factors will not cause this situation to occur.

36. **(c)** Control by noise factor interactions are the key to robustness and the experiment must be structured to find these interactions. Option (a) is therefore unsatisfactory. Option (b) is satisfactory but requires more trials than necessary. Option (c) is the best because it allows the determination of control by joint noise interactions in the least number of trials. Option (d) will result in a very large experiment, particularly when the quadratic effects are not expected to be large. Option (e) is not satisfactory because it will not permit all control by noise interactions to be determined (there are potentially 10 such interactions if the two noise factors are considered separately) without confounding.

37. **(b)** Both A and T are noise factors (A is also a control factor). Thus, the control by noise interactions AB, AT, and BT will be helpful in achieving robustness. Also, the quadratic effect of A will be useful because the level of A could be selected to minimize the slope of A versus the response relationship. The quadratic effect of B will not be useful in accomplishing robustness since B is well controlled.

38. **(b)** Pull strength is an example of a larger-the-better characteristic. The correct signal-to-noise ratio to use is:

$$\text{S/N ratio} = -10\log\left(\frac{1}{n}\sum\frac{1}{y_i^2}\right)$$

Maximizing this ratio implies large pull strength. The pull strength values are squared in the formula above because that minimizes the quadratic loss.

39. **(a)** The missing value is not of much concern for the replicated two-factor factorial since the remaining replicate is available under that experimental condition and all effects can be estimated. With a three-factor unreplicated factorial, one of the effects cannot be estimated. This may be taken to be the three-factor interaction, which is usually small. With unreplicated four factors in eight trials, we will not be able to estimate one two-factor interaction, and with seven factors in eight trials, we will not be able to estimate one main effect. The correct answer is (a) with (e) being a close second.

40. **(d)** The right sequence is to select responses first; noise factors next, and control factors last. The selection of noise factors is a function of knowing which outputs (responses) are of interest. Similarly, the selected control factors will depend upon the output of interest, and control factors should also be selected with the possibility that they will interact with the noise factors, making robustness possible.

41. **(e)** Noise factors can be external to the product or the process, internal to the product and the process, or factors that cause product degradation. Replication is one way to assess the impact of internal and some external noise factors. So the answer is all of the above.

42. **(e)** As the product characteristic deviates from target, product quality degrades. The losses sustained by the customers increase more than proportionately as the product deviates from target. The quadratic loss function is the simplest function that describes this relationship between deviation from target and customer loss. The customer loss is proportional to the square of the deviation from target. Thus, the ideas that all products inside a specification are equally good, or that a product just outside the specification is very bad and should be rejected, and another product just inside the specification is good and should be accepted are incorrect.

43. **(c)** The empirical specification is expressed by the tolerance interval $\bar{x} \pm ks$, where the value of k, tabulated in the Appendix depends upon the percent confidence (95 percent), percentage of the population to be enclosed inside the specification (99 percent), and the sample size (13). This value of k is 4.044. Hence, the specification is $5 \pm (4.044)(0.5)$ or 5 ± 2. It should be noted that the empirical approach to setting specifications is not as satisfactory as the functional and minimum life cycle cost approaches to setting specifications.

44. **(d)** Worst-case specifications are derived under the assumption that the characteristic can take the worst possible value with 100 percent probability. Each of the three-stacked parts could simultaneously be at their largest dimension or at their smallest dimension. Therefore, the worst-case specification for stack height is 6 ± 0.3. Worst-case specifications are overly conservative.

45. (c) Statistical specification assumes that the characteristic has a normal distribution with the specified standard deviation. The mean is assumed to be perfectly centered on target. It banks upon the characteristic taking values according to this probability distribution. The assumption that the mean is perfectly centered is unrealistic; therefore, statistical specifications are too optimistic. In our example, two similar parts are stacked to produce a stack height with a target of 10; hence, the target for each part is 5. Since the variance for stack height is 1.0, the variance for each part must be 0.5. Thus, the standard deviation for each part is 0.7. The statistical specification for the part is $5.0 < 0.7$. The 3-sigma limits for each part are 5.0 ± 2.1.

46. (a) The worst-case specification for each part is 5 ± 1.5. Note that this specification is narrower than the 3-sigma limits in the above example. This means that for a similar specification on output, statistical specifications on inputs are wider than the worst-case specifications.

47. (b) Customer loss $= k(y - T)^2$ where y is the voltage of a specific converter and T is the target voltage. A deviation of 10 volts from target causes a customer loss of $100. Hence, $k = 1$. The manufacturer can adjust the voltage for a cost of $4. From the customer loss equation, a voltage deviation from target of >2 volts will cause the customer to lose more money than it costs the producer to fix the product. Therefore, the producer's voltage specification should be $T \pm 2$ volts.

48. (b) In this situation, a functional—worst-case approach should be used to set specifications. The reason is that a hospital may use one machine for an extended period of time. The producer cannot rely upon probabilities of occurrence, each machine has to function satisfactorily even at the edge of the specification.

49. (d) $P_p =$ specification width/$6\sigma_{\text{total}} = 1.33$ and for a specification width of 8, $\sigma_{\text{total}} = 1$. Therefore, $\sigma_{\text{total}}^2 = 1$ and this total variance is equally divided between R&D and manufacturing because the job of these two organizations is deemed to be equally difficult. Thus, $\sigma_{\text{R\&D}}^2 = \sigma_{\text{manufacturing}}^2 = 0.5$. The allocated standard deviations for R&D and manufacturing are equal to 0.7 each.

50. (e) Worst-case specifications can be overly conservative. Statistical specifications can be overly optimistic. This means that for a given specification on output, the worst-case specifications for inputs are tighter. For a given specification for inputs, the worst-case specification for output is wider.

51. (c) In this case, it is the percentage specification for Y that gets allocated to A and B. The percentage specification for Y is ± 10 percent. Option (c) is the correct answer because, in this case, the percentage specification for A is ± 2 percent and the percentage specification for B is ± 8 percent. The two percentage specifications add up to 10 percent. For all other options, this is not the case.

52. (d) The acceptable quality level (AQL) is measured in percent defective units. It is the largest percent defective that is acceptable as a process average.

53. (b) Just because a particular lot is accepted by an acceptance sampling plan does not mean that the accepted lot is a good lot. Every acceptance sampling plan has a certain probability of accepting lots with various fraction defectives. Thus, a bad lot may also get accepted, although usually the probability of such an event is small. Similarly, a good lot may occasionally get rejected. If the manufacturing process is perfectly stable, every lot produced will have essentially the same fraction defective. If these lots are subjected to an acceptance sampling plan, some will get accepted and the remaining will get rejected. Accepting and rejecting the same quality lots is counterproductive. Accept-on-zero plans have an acceptance number equal to zero. The fact that the sample did not contain a defective item does not mean that all accepted lots have zero defectives. Both AQL and RQL should be jointly decided by producers and customers. For example, AQL is often close to the true quality produced by the producer and accepted by the customer. If the producer were to arbitrarily select a large value of AQL, the customer will end up with large fraction defective.

54. (e) If zero defectives are found in a sample of size n, then we are 95 percent sure that the maximum value of fraction defective is $300/n$ percent. Since zero defectives were found in a sample of size 50, we are 95 percent sure that the true percent defective is less than 6 percent.

55. (c) The RQL to be used in validation should be equal to the AQL to be used in manufacturing. This is so because the purpose of validation is to demonstrate that the fraction defective in routine manufacturing is less than the AQL to be used in manufacturing.

56. (c) If the lot contains 1 percent defectives, then the probability of randomly picking a single acceptable product is 0.99. The lot will be accepted if 10 successively picked products are all acceptable. The probability of such an event is $0.99^{10} = 0.90 = 90$ percent.

57. (d) A critical characteristic is one which, if outside specification, will cause a safety concern. It should have a small AQL and a small RQL. If several lots fail, it means that the production quality has degraded and tightened inspection should be instituted by reducing the AQL and RQL of the sampling plan. If CQL is less than AQL, few rather than many lots will fail. Finally, industry practice can be and often is used to select AQL and RQL.

58. **(e)** If an acceptance sampling plan has been proposed by a customer, the first action a supplier should take is to calculate the probability of routine production being rejected by this sampling plan. This is done by calculating the fraction defective in routine production and computing the corresponding risk of rejection based upon the operating characteristic curve of the sampling plan. Appropriate other actions will follow based upon whether the risk of lot rejection is acceptable or not.

59. **(e)** For the given operating characteristic curve, AQL, corresponding to a 5 percent risk of lot rejection is 0.6 percent. The RQL, corresponding to a 10 percent risk of lot acceptance is 5 percent. A lot with 3 percent defectives will be rejected 60 percent of the time and a lot with 6 percent defectives will be accepted 5 percent of the time. If CQL exceeds AQL, then the probability of rejecting routine production will exceed 5 percent and will become unacceptable.

60. **(d)** If the lot contains 5 percent defectives, the probability of accepting the lot on the first sample is $0.95^{10} = 60$ percent. The probability of rejecting the lot on the first sample is 40 percent. The probability of rejecting the lot on the first sample, and then accepting it on the second sample is $(40\%)\,(60\%) = 24$ percent. Hence, the total probability of accepting the lot under this sampling scheme is 84 percent.

61. **(d)** A double sampling plan usually has a smaller expected sample size compared with a single sample plan, although a double sampling plan is more difficult to administer. A variable sampling plan has a smaller sample size than an attribute sampling plan for the same operating characteristic curve, although variable measurements are more expensive than attribute measurements. It is not true that a variable sampling plan will provide greater protection than an attribute sampling plan. The protection provided by a sampling plan, as quantified by the operating characteristic curve, is a function of the chart parameters such as the sample size. Plans with the same operating characteristic curve provide the same protection.

62. **(d)** For an $\bar{X}$ chart, the centerline is often taken to be the grand average of data. The control limits are $3\sigma/\sqrt{n}$ away from the centerline. The reason to divide by $\sqrt{n}$ is because the standard deviation of the average reduces by a factor of $\sqrt{n}$. The reason 3 is used is because we accept a false alarm probability of 3/1000 corresponding to the 3 standard deviation limits. The data are assumed to be normally distributed. The standard deviation σ is the short-term standard

deviation taken to be the *within-subgroup* standard deviation, *not* the between-subgroup standard deviation.

63. **(d)** Process mean has no bearing on determining the sampling interval. If the process shifts often, if the cost of bad product is high, if the capability index is small, and if the cost of sampling and testing is small, the sampling interval will be smaller.

64. **(b)** The risk of false alarm is always set at 3/1000 because we use 3-sigma limit charts. Other listed factors are important in determining the subgroup size. As subgroup size increases, it becomes easier to catch process shifts, the cost of sampling and testing increases, but the cost of bad product reduces. If the capability index is high, the benefits of increasing the subgroup size reduce because the probability of producing bad product reduces.

65. **(d)** For a chart of averages, the control limits are $3\sigma/\sqrt{n}$ away from the centerline. The current chart has a subgroup size of 4, a centerline of 10, and upper and lower control limits of 12 and 8, respectively. If the subgroup size is changed to 1, the distance between the centerline and control limits will double. The new limits will be 14 and 6, respectively.

66. **(c)** A control chart should be redesigned when a deliberate process change has been made and the chart indicates that the process has changed. It may also be redesigned when sufficient data are collected if the original design of the chart was based upon inadequate amount of data.

67. **(a)** For a control chart, β risk refers to the probability of not detecting a shift in mean on the first subgroup taken immediately after the shift has occurred. The α risk refers to the probability of false alarm, namely, the probability of concluding that the process is out-of-control when it is not.

68. **(b)** The out-of-control rules generally have a probability of 3/1000 to 5/1000 of occurring by pure chance. In other words, they are rare events if the process is in control. The probability that four successive points will fall below the center line is $(1/2)^4 = 0.0625$ or over 60/1000. The correct out-of-control rule is *eight points in a row below the centerline* and the probability of this happening by pure chance is 4/1000.

69. **(a)** Since C_p and C_{pk} are equal, the process is centered. Since P_{pk} is much smaller than C_{pk}, the process needs to be stabilized. The process does not need to be fundamentally changed. Once the process is stable, it will have a satisfactory capability.

70. **(d)** The difference between the C_{pk} index and the P_{pk} index has to do with stability. The C_{pk} index assumes the process to be stable, the P_{pk} index does not.

71.   **(b)** With reference to the upper specification limit, the P_{pk} index is defined as:

$$P_{pk} = \frac{\text{upper specification} - \text{mean}}{3\sigma_{\text{total}}}$$

The only way the index will become negative is if the mean exceeds the upper specification limit. More broadly, if the mean is outside the specification the index will be negative.

72.   **(e)** If the process is improved by centering it and reducing its variability; both P_p and P_{pk} will reduce and will become equal to each other.

73.   **(a)** The C_{pk} index is defined to be the smaller of

$$C_{pk} = \frac{\text{upper specification} - \text{mean}}{3\sigma_{\text{short}}} \quad \text{or} \quad \frac{\text{mean} - \text{lower specification}}{3\sigma_{\text{short}}}$$

Substituting upper specification $= 10$, mean $= 8$, and $\sigma_{\text{short}} = 2$, we have $C_{pk} = 0.33$.

74.   **(b)** Stability and capability of a process are two separate concepts. Stability means that the process does not change over time. It is not necessary to know the specifications to assess stability. Capability is the ability of the process to produce products within specifications. Thus, a stable process may or may not be capable.

75.   **(e)** The purpose of variance components analysis is to partition total variance into variance due to various causes of variability. Variance components analysis has many applications including determining the variance due to raw material batch-to-batch variability, determining within- and between-subgroup variances and determining the repeatability and reproducibility of a measurement system. It is usually not used to assess if there is a linear time trend to the data. Regression analysis is a better tool for such an assessment.

76.   **(a)** The within-batch variance for the first batch is 0.5 and it is the same for the second batch. Therefore, a pooled estimate of within-batch variance is 0.5. The averages for the first and second batch are 8.5 and 10.5, respectively. The variance computed from these two batch averages is 2.0. This variance is equal to the batch-to-batch variance plus half of within-batch variance. Hence, the batch-to-batch variance is 1.75.

77.   **(e)** Note that the variance within subgroup, which is due to common causes of variability, is much larger than the variance between subgroups, which is due to special causes that occur over time. Therefore, to significantly reduce variability, it is the common cause variability that must be reduced. Among the actions listed, evaluating the measurement system variability is the only cause that may reduce the common cause variation. All other listed causes have to do with between-subgroup variation.

78. **(e)** For a centered process, the average quadratic loss is directly proportional to the variance. If the standard deviation is reduced by a factor of two, variance reduces by a factor of four and the average quadratic loss will be $0.5 million.

79. **(b)** To determine the variance component due to humidity, a very low humidity day and a very high humidity day were purposefully selected for experimentation. These two humidity values are not a random drawing from the distribution of humidity throughout the year. They overestimate the variance of humidity, and thereby overestimate the variance contribution due to humidity.

80. **(a)** The P_{pk} index is

$$P_{pk} = \frac{\text{mean} - \text{lower specification limit}}{3\sigma_{\text{total}}}$$

The index will improve if the mean is centered and/or if the total standard deviation is reduced. The total variance is 7; hence, the total standard deviation is 2.645. The current $P_{pk} = 0.25$. If the only action taken is to center the mean, the index will become 0.5. The largest variance component is due to manufacturing. If the only action taken is to eliminate manufacturing variability, the total standard deviation will be 1.732, and the P_{pk} index will be 0.38. Centering the mean is the single action that will improve the index the most.

81. **(e)** Since all the machines that the company uses are in the study, machines are a fixed factor. Products are nested inside the machines because the 100 products for each machine are different from machine to machine.

82. **(c)** The within-subgroup or the common cause variance is equal to 1.0. The focus must be on reducing this common cause variance, not the special cause variance, which is substantially smaller. Whether the total variance is too large or not cannot be determined without knowing the specification for the characteristic under consideration. Twenty or more degrees of freedom are reasonable to estimate a variance component, the larger the degrees of freedom the better.

83. **(b)** Since the same six people visited the same four restaurants, people and restaurants are crossed. Since each person ordered their own separate pizzas, pizzas are nested inside the people and restaurants.

84. **(d)** The total standard deviation of the measurement system should be less than $W/10$, where W is the half-width of a two-sided specification or the

distance between the process mean and the single-sided specification limit. The main reason for this is the fact that we are unwilling to allocate a very large proportion of the specification to the measurement system variability. There are many other sources of variability, such as manufacturing variability, that are often much larger than measurement system variance. A large portion of the specification needs to be reserved for causes other than the measurement system.

85. **(b)** When a measurement system is used for improvement purposes, it should be able to distinguish products into many statistically distinct categories. This requires the measurement standard deviation to be less than a third of the total standard deviation. The total standard deviation includes the product-to-product variability and the measurement variability. When the above condition is satisfied, the measurement system is able to classify products into at least four distinct categories, making it significantly better than an attribute measurement system, which classifies products into only two categories.

86. **(c)** For nondestructive tests, the true measurement variability can be determined by repeated testing of the same product. For destructive tests, the same product cannot be tested twice. The so-called measurement variability for destructive tests also includes the short-term product-to-product variability. This is the main reason why the acceptance criteria for destructive tests are wider.

87. **(c)** A large slope indicates high sensitivity and a small standard deviation means small variability. The measurement system that has high sensitivity coupled with small variability is the best. Therefore, it is the ratio of slope to standard deviation that matters and a measurement system with the highest ratio is the best.

88. **(a)** The duplicates are nested inside the operators and since two *different* operators measure each product, the operators are nested inside the products. If the same two operators had measured all products, then the operators and products would have been crossed.

89. **(c)** The fact that there is a large interaction between products and operators means that there are differences in the way the operators measure each product. The right course of action is to train the operators to measure alike. This will require understanding the key steps in the measurement process. Perhaps it is the way the sample is prepared if sample preparation is a part of the measurement process, or it is the way the product is aligned before taking measurements.

90. **(e)** Every factor produces some effect, however small. Some of these effects are detected as statistically significant depending upon the sample size. Large

sample sizes will detect very small effects. Just because an effect is statistically significant does not mean that it is practically important. From the information given in this example, it is not possible to judge the size of a practically important effect. Therefore, conclusions regarding robustness of the measurement system or whether the specification for factor A should be constrained cannot be made. Similarly, it is not possible to judge whether the measurement system is acceptable for improvement or specification-related purposes because this assessment depends upon knowing the total standard deviation and the specification, respectively—information that is not provided here.

91. **(a)** Note that P_{pk} index depends upon the total standard deviation and is a function of the measurement and product standard deviations that add together as follows:

$$\sigma_{\text{total}}^2 = \sigma_{\text{measurement}}^2 + \sigma_{\text{product}}^2$$

Since all suppliers have the same P_{pk}, they have the same total variance. The supplier with the largest measurement variance has the smallest true product-to-product variance and it is the product-to-product variability that the customer cares about.

92. **(c)** If the total variance is taken to be 100, then the measurement variance is 10. This means that the ratio $\sigma_{\text{total}}/\sigma_{\text{measurement}} = \sqrt{100}/\sqrt{10} > 3.0$ and the measurement system is satisfactory for improvement purposes. Whether the measurement system is satisfactory for specification purposes or not requires the knowledge of specification. There is operator-to-operator variability, not because this variance component exceeds instrument variability, but because this variance component is the largest. Whether the operators must be retrained depends upon whether the measurement system is satisfactory for the intended purpose or not. Also, whether the product-to-product variability is too large or not can only be judged in the context of specification, which has not been provided in this example.

93. **(d)** Adjusting a measurement system is different from simply conducting a calibration experiment. A measurement system should be adjusted when a control chart of measurements of a standard (or an unchanging product) over time produces an out-of-control signal. Otherwise, adjustments will be made based upon noise, and the measurement variability will in-fact increase.

94. **(b)** The volume of injected fluid is equal to the cross-sectional area of the syringe multiplied by the stroke length minus the volume of back flow. This is given by

$$\text{Injected volume} = \pi R^2 L - B$$

Knowing mechanistic models similar to this can help dramatically in setting specifications.

95. **(b)** For multiplicative models, CV^2's add. In this case,

$$CV_y^2 = CV_A^2 + \frac{1}{4}CV_B^2$$

Since $CV_A = 10$ percent and $CV_B = 25$ percent, upon substitution, $CV_y = 16$ percent.

96. **(e)** Both A and B influence the output. Since R^2 has turned out to be small, it is possible that another factor is influencing the output or that the measurement variance is large. It is not necessary for the measurement variance to be small for the effects of A and B to be statistically significant. This is because statistical significance is also a function of the number of observations. With a very large number of observations, even if the variability is large, small effects can be detected as statistically significant.

97. **(c)** In linear regression, R^2 (actually adjusted R^2) is approximately given by

$$R^2 = 1 - \frac{\sigma_e^2}{\sigma_Y^2}$$

where σ_e^2 is residual variance and σ_Y^2 is the variance of output. If the range of X is large, Y will vary over a wider range, σ_Y^2 will be large and R^2 will increase. If the measurement error is small, σ_e^2 will reduce and R^2 will increase. If no other factor influences Y, then σ_e^2 will reduce and R^2 will increase. If the range of X is small, Y will vary over a narrow range, σ_Y^2 will be small, and R^2 will reduce.

98. **(e)** All the statements are correct. If the power function form of the relationship is unknown, the designed experiment will fit an additive linear model (corresponding to the Taylor series expansion of the power function) and in doing so, the experiment will become unnecessarily large and the fitted equation will only be approximate, resulting in larger prediction errors. The scientific understanding gained from this approximate and unnecessarily complex equation will be less than satisfactory. For these various reasons, it is important to attempt to postulate the likely form of the relationship before designing and conducting experiments.

99. **(c)** If the time series is positively autocorrelated, it will exhibit a cyclic behavior. The control chart run rules, which are based upon the assumption that successive observations are uncorrelated, will not apply. Because successive observations are related to each other, future values can be predicted within certain bounds using time series models. Based upon large number of observations, the mean can be estimated.

100. **(e)** The fitted straight line may go through the origin because the confidence interval for the intercept includes zero. The slope is statistically significant because the confidence interval for the slope does not include zero. $R^2 = 88$ percent means that the input factor explains 88 percent of the variance of output. The 95 percent prediction interval for predicted values will be approximately twice the prediction error standard deviation, namely, ± 3.0. Whether the fitted model is very useful or not depends upon the practical use the model is intended for. If the requirement was to make predictions within ± 1, the model does not meet the requirement.

APPENDIX

TABLES

Industrial Statistics: Practical Methods and Guidance for Improved Performance   By Anand M. Joglekar
Copyright © 2010 John Wiley & Sons, Inc.

TABLE A.1 **Tail Area of Unit Normal Distribution**

z	0.00	0.01	0.02	0.03	0.04	0.05	0.06	0.07	0.08	0.09
0.0	0.5000	0.4960	0.4920	0.4880	0.4840	0.4801	0.4761	0.4721	0.4681	0.4641
0.1	0.4602	0.4562	0.4522	0.4483	0.4443	0.4404	0.4364	0.4325	0.4286	0.4247
0.2	0.4207	0.4168	0.4129	0.4090	0.4052	0.4013	0.3974	0.3936	0.3897	0.3859
0.3	0.3821	0.3783	0.3745	0.3707	0.3669	0.3632	0.3594	0.3557	0.3520	0.3483
0.4	0.3446	0.3409	0.3372	0.3336	0.3300	0.3264	0.3228	0.3192	0.3156	0.3121
0.5	0.3085	0.3050	0.3015	0.2981	0.2946	0.2912	0.2877	0.2843	0.2810	0.2776
0.6	0.2743	0.2709	0.2676	0.2643	0.2611	0.2578	0.2546	0.2514	0.2483	0.2451
0.7	0.2420	0.2389	0.2358	0.2327	0.2296	0.2266	0.2236	0.2206	0.2177	0.2148
0.8	0.2119	0.2090	0.2061	0.2033	0.2005	0.1977	0.1949	0.1922	0.1894	0.1867
0.9	0.1841	0.1814	0.1788	0.1762	0.1736	0.1711	0.1685	0.1660	0.1635	0.1611
1.0	0.1587	0.1562	0.1539	0.1515	0.1492	0.1469	0.1446	0.1423	0.1401	0.1379
1.1	0.1357	0.1335	0.1314	0.1292	0.1271	0.1251	0.1230	0.1210	0.1190	0.1170
1.2	0.1151	0.1131	0.1112	0.1093	0.1075	0.1056	0.1038	0.1020	0.1003	0.0985
1.3	0.0968	0.0951	0.0934	0.0918	0.0901	0.0885	0.0869	0.0853	0.0838	0.0823
1.4	0.0808	0.0793	0.0778	0.0764	0.0749	0.0735	0.0721	0.0708	0.0694	0.0681
1.5	0.0668	0.0655	0.0643	0.0630	0.0618	0.0606	0.0594	0.0582	0.0571	0.0559
1.6	0.0548	0.0537	0.0526	0.0516	0.0505	0.0495	0.0485	0.0475	0.0465	0.0455
1.7	0.0446	0.0436	0.0427	0.0418	0.0409	0.0401	0.0392	0.0384	0.0375	0.0367
1.8	0.0359	0.0351	0.0344	0.0336	0.0329	0.0322	0.0314	0.0307	0.0301	0.0294
1.9	0.0287	0.0281	0.0274	0.0268	0.0262	0.0256	0.0250	0.0244	0.0239	0.0233
2.0	0.0228	0.0222	0.0217	0.0212	0.0207	0.0202	0.0197	0.0192	0.0188	0.0183
2.1	0.0179	0.0174	0.0170	0.0166	0.0162	0.0158	0.0154	0.0150	0.0146	0.0143
2.2	0.0139	0.0136	0.0132	0.0129	0.0125	0.0122	0.0119	0.0116	0.0113	0.0110
2.3	0.0107	0.0104	0.0102	0.0099	0.0096	0.0094	0.0091	0.0089	0.0087	0.0084
2.4	0.0082	0.0080	0.0078	0.0075	0.0073	0.0071	0.0069	0.0068	0.0066	0.0064
2.5	0.0062	0.0060	0.0059	0.0057	0.0055	0.0054	0.0052	0.0051	0.0049	0.0048
2.6	0.0047	0.0045	0.0044	0.0043	0.0041	0.0040	0.0039	0.0038	0.0037	0.0036
2.7	0.0035	0.0034	0.0033	0.0032	0.0031	0.0030	0.0029	0.0028	0.0027	0.0026
2.8	0.0026	0.0025	0.0024	0.0023	0.0023	0.0022	0.0021	0.0021	0.0020	0.0019
2.9	0.0019	0.0018	0.0018	0.0017	0.0016	0.0016	0.0015	0.0015	0.0014	0.0014
3.0	0.0013	0.0013	0.0013	0.0012	0.0012	0.0011	0.0011	0.0011	0.0010	0.0010
3.1	0.0010	0.0009	0.0009	0.0009	0.0008	0.0008	0.0008	0.0008	0.0007	0.0007
3.2	0.0007	0.0007	0.0006	0.0006	0.0006	0.0006	0.0006	0.0005	0.0005	0.0005
3.3	0.0005	0.0005	0.0005	0.0004	0.0004	0.0004	0.0004	0.0004	0.0004	0.0003
3.4	0.0003	0.0003	0.0003	0.0003	0.0003	0.0003	0.0003	0.0003	0.0003	0.0002
3.5	0.0002	0.0002	0.0002	0.0002	0.0002	0.0002	0.0002	0.0002	0.0002	0.0002
3.6	0.0002	0.0002	0.0001	0.0001	0.0001	0.0001	0.0001	0.0001	0.0001	0.0001
3.7	0.0001	0.0001	0.0001	0.0001	0.0001	0.0001	0.0001	0.0001	0.0001	0.0001
3.8	0.0001	0.0001	0.0001	0.0001	0.0001	0.0001	0.0001	0.0001	0.0001	0.0001
3.9	0.0000	0.0000	0.0000	0.0000	0.0000	0.0000	0.0000	0.0000	0.0000	0.0000

TABLE A.2 Probability Points of the *t*-Distribution with *v* Degrees of Freedom

					Tail Area Probability					
v	0.4	0.25	0.1	0.05	0.025	0.01	0.005	0.0025	0.001	0.0005
1	0.325	1.000	3.078	6.314	12.706	31.821	63.657	127.32	318.31	636.62
2	0.289	0.816	1.886	2.920	4.303	6.965	9.925	14.089	22.326	31.598
3	0.277	0.765	1.638	2.353	3.182	4.541	5.841	7.453	10.213	12.924
4	0.271	0.741	1.533	2.132	2.776	3.747	4.604	5.598	7.173	8.610
5	0.267	0.727	1.476	2.015	2.571	3.365	4.032	4.773	5.893	6.869
6	0.265	0.718	1.440	1.943	2.447	3.143	3.707	4.317	5.208	5.959
7	0.263	0.711	1.415	1.895	2.365	2.998	3.499	4.029	4.785	5.408
8	0.262	0.706	1.397	1.860	2.306	2.896	3.355	3.833	4.501	5.041
9	0.261	0.703	1.383	1.833	2.262	2.821	3.250	3.690	4.297	4.781
10	0.260	0.700	1.372	1.812	2.228	2.764	3.169	3.581	4.144	4.587
11	0.260	0.697	1.363	1.796	2.201	2.718	3.106	3.497	4.025	4.437
12	0.259	0.695	1.356	1.782	2.179	2.681	3.055	3.428	3.930	4.318
13	0.259	0.694	1.350	1.771	2.160	2.650	3.012	3.372	3.852	4.221
14	0.258	0.692	1.345	1.761	2.145	2.624	2.977	3.326	3.787	4.140
15	0.258	0.691	1.341	1.753	2.131	2.602	2.947	3.286	3.733	4.073
16	0.258	0.690	1.337	1.746	2.120	2.583	2.921	3.252	3.686	4.015
17	0.257	0.689	1.333	1.740	2.110	2.567	2.898	3.222	3.646	3.965
18	0.257	0.688	1.330	1.734	2.101	2.552	2.878	3.197	3.610	3.922
19	0.257	0.688	1.328	1.729	2.093	2.539	2.861	3.174	3.579	3.883
20	0.257	0.687	1.325	1.725	2.086	2.528	2.845	3.153	3.552	3.850
21	0.257	0.686	1.323	1.721	2.080	2.518	2.831	3.135	3.527	3.819
22	0.256	0.686	1.321	1.717	2.074	2.508	2.819	3.119	3.505	3.792
23	0.256	0.685	1.319	1.714	2.069	2.500	2.807	3.104	3.485	3.767
24	0.256	0.685	1.318	1.711	2.064	2.492	2.797	3.091	3.467	3.745
25	0.256	0.684	1.316	1.708	2.060	2.485	2.787	3.078	3.450	3.725
26	0.256	0.684	1.315	1.706	2.056	2.479	2.779	3.067	3.435	3.707
27	0.256	0.684	1.314	1.703	2.052	2.473	2.771	3.057	3.421	3.690
28	0.256	0.683	1.313	1.701	2.048	2.467	2.763	3.047	3.408	3.674
29	0.256	0.683	1.311	1.699	2.045	2.462	2.756	3.038	3.396	3.659
30	0.256	0.683	1.310	1.697	2.042	2.457	2.750	3.030	3.385	3.646
40	0.255	0.681	1.303	1.684	2.021	2.423	2.704	2.971	3.307	3.551
60	0.254	0.679	1.296	1.671	2.000	2.390	2.660	2.915	2.232	3.460
120	2.254	0.677	1.289	1.658	1.980	2.358	2.617	2.860	3.160	3.373
∞	0.253	0.674	1.282	1.645	1.960	2.326	2.576	2.807	3.090	3.291

Source: From E. S. Pearson and H. O. Hartley (Eds.) (1958), *Biometrika Tables for Statisticians*, Vol. 1, used by permission of Oxford University Press.

TABLE A.3 **Probability Points of the χ^2 Distribution with v Degrees of Freedom**

| v | \multicolumn{9}{c}{Tail Area Probability} |
	0.995	0.990	0.975	0.950	0.500	0.050	0.025	0.010	0.005
1	0.00+	0.00+	0.00+	0.00+	0.45	3.84	5.02	6.63	7.88
2	0.01	0.02	0.05	0.10	1.39	5.99	7.38	9.21	10.60
3	0.07	0.11	0.22	0.35	2.37	7.81	9.35	11.34	12.84
4	0.21	0.30	0.48	0.71	3.36	9.49	11.14	13.28	14.86
5	0.41	0.55	0.83	1.15	4.35	11.07	12.83	15.09	16.75
6	0.68	0.87	1.24	1.64	5.35	12.59	14.45	16.81	18.55
7	0.99	1.24	1.69	2.17	6.35	14.07	16.01	18.48	20.28
8	1.34	1.65	2.18	2.73	7.34	15.51	17.53	20.09	21.96
9	1.73	2.09	2.70	3.33	8.34	16.92	19.02	21.67	23.59
10	2.16	2.56	3.25	3.94	9.34	18.31	20.48	23.21	25.19
11	2.60	3.05	3.82	4.57	10.34	19.68	21.92	24.72	26.76
12	3.07	3.57	4.40	5.23	11.34	21.03	23.34	26.22	28.30
13	3.57	4.11	5.01	5.89	12.34	22.36	24.74	27.69	29.82
14	4.07	4.66	5.63	6.57	13.34	23.68	26.12	29.14	31.32
15	4.60	5.23	6.27	7.26	14.34	25.00	27.49	30.58	32.80
16	5.14	5.81	6.91	7.96	15.34	26.30	28.85	32.00	34.27
17	5.70	6.41	7.56	8.67	16.34	27.59	30.19	33.41	35.72
18	6.26	7.01	8.23	9.39	17.34	28.87	31.53	34.81	37.16
19	6.84	7.63	8.91	10.12	18.34	30.14	32.85	36.19	38.58
20	7.43	8.26	9.59	10.85	19.34	31.41	34.17	37.57	40.00
25	10.52	11.52	13.12	14.61	24.34	37.65	40.65	44.31	46.93
30	13.79	14.95	16.79	18.49	29.34	43.77	46.98	50.89	53.67
40	20.71	22.16	24.43	26.51	39.34	55.76	59.34	63.69	66.77
50	27.99	29.71	32.36	34.76	49.33	67.50	71.42	76.15	79.49
60	35.53	37.48	40.48	43.19	59.33	79.08	83.30	88.38	91.95
70	43.28	45.44	48.76	51.74	69.33	90.53	95.02	100.42	104.22
80	51.17	53.54	57.15	60.39	79.33	101.88	106.63	112.33	116.32
90	59.20	61.75	65.65	69.13	89.33	113.14	118.14	124.12	128.30
100	67.33	70.06	74.22	77.93	99.33	124.34	129.56	135.81	140.17

Source: From E. S. Pearson and H. O. Hartley (Eds.) (1966), *Biometrika Tables for Statisticians*, Vol. 1, used by permission of Oxford University Press.

TABLE A.4   *k* Values for Two-Sided Normal Tolerance Limits

	90 Percent Confidence that Percentage of Population Between Limits Is			95 Percent Confidence that Percentage of Population Between Limits Is			99 Percent Confidence that Percentage of Population Between Limits Is		
n	90%	95%	99%	90%	95%	99%	90%	95%	99%
2	15.98	18.80	24.17	32.02	37.67	48.43	160.2	188.5	242.3
3	5.847	6.919	8.974	8.380	9.916	12.86	18.93	22.40	29.06
4	4.166	4.943	6.440	5.369	6.370	8.299	9.398	11.15	14.53
5	3.494	4.152	5.423	4.275	5.079	6.634	6.612	7.855	10.26
6	3.131	3.723	4.870	3.712	4.414	5.775	5.337	6.345	8.301
7	2.902	3.452	4.521	3.369	4.007	5.248	4.613	5.448	7.187
8	2.743	3.264	4.278	3.136	3.732	4.891	4.147	4.936	6.468
9	2.626	3.125	4.098	2.967	3.532	4.631	3.822	4.550	5.966
10	2.535	3.018	3.959	2.829	3.379	4.433	3.582	4.265	5.594
11	2.463	2.933	3.849	2.737	3.259	4.277	3.397	4.045	5.308
12	2.404	2.863	3.758	2.655	3.162	4.150	3.250	3.870	5.079
13	2.355	2.805	3.682	2.587	3.081	4.044	3.130	3.727	4.893
14	2.314	2.756	3.618	2.529	3.012	3.955	3.029	3.608	4.737
15	2.278	2.713	3.562	2.480	2.954	3.878	2.945	3.507	4.605
16	2.246	2.676	3.514	2.437	2.903	3.812	2.872	3.421	4.492
17	2.219	2.643	3.471	2.400	2.858	3.754	2.808	3.345	4.393
18	2.194	2.614	3.433	2.366	2.819	3.702	2.753	3.279	4.307
19	2.172	2.588	3.399	2.337	2.784	3.656	2.703	3.221	4.230
20	2.152	2.564	3.368	2.310	2.752	3.615	2.659	3.168	4.161
21	2.135	2.543	3.340	2.286	2.723	3.577	2.620	3.121	4.100
22	2.118	2.524	3.315	2.264	2.697	3.543	2.584	3.078	4.044
23	2.103	2.506	3.292	2.244	2.673	3.512	2.551	3.040	3.993
24	2.089	2.489	3.270	2.225	2.651	3.483	2.522	3.004	3.947
25	2.077	2.474	3.251	2.208	2.631	3.457	2.494	2.972	3.904
26	2.065	2.460	3.232	2.193	2.612	3.432	2.469	2.941	3.865
27	2.054	2.447	3.215	2.178	2.595	3.409	2.446	2.914	3.828
28	2.044	2.435	3.199	2.164	2.579	3.388	2.424	2.888	3.794
29	2.034	2.424	3.184	2.152	2.554	3.368	2.404	2.864	3.763
30	2.025	2.413	3.170	2.140	2.549	3.350	2.385	2.841	3.733
35	1.988	2.368	3.112	2.090	2.490	3.272	2.306	2.748	3.611
40	1.959	2.334	3.066	2.052	2.445	3.213	2.247	2.677	3.518
50	1.916	2.284	3.001	1.996	2.379	3.126	2.162	2.576	3.385
60	1.887	2.248	2.955	1.958	2.333	3.066	2.103	2.506	3.293
80	1.848	2.202	2.894	1.907	2.272	2.986	2.026	2.414	3.173
100	1.822	2.172	2.854	1.874	2.233	2.934	1.977	2.355	3.096
200	1.764	2.102	2.762	1.798	2.143	2.816	1.865	2.222	2.921
500	1.717	2.046	2.689	1.737	2.070	2.721	1.777	2.117	2.783
1000	1.695	2.019	2.654	1.709	2.036	2.676	1.736	2.068	2.718
∞	1.645	1.960	2.576	1.645	1.960	2.576	1.645	1.960	2.576

Source: From D. C. Montgomery (1985), *Introduction to Statistical Quality Control*, used by permission of John Wiley & Sons, Inc.

TABLE A.5 *k* **Values for One-Sided Normal Tolerance Limits**

	90 Percent Confidence that Percentage of Population is Below (Above) Limit Is			95 Percent Confidence that Percentage of Population is Below (Above) Limit Is			99 Percent Confidence that Percentage of Population is Below (Above) Limit Is		
n	90%	95%	99%	90%	95%	99%	90%	95%	99%
3	4.258	5.310	7.340	6.158	7.655	10.552			
4	3.187	3.957	5.437	4.163	5.145	7.042			
5	2.742	3.400	4.666	3.407	4.202	5.741			
6	2.494	3.091	4.242	3.006	3.707	5.062	4.408	5.409	7.334
7	2.333	2.894	3.972	2.755	3.399	4.641	3.856	4.730	6.411
8	2.219	2.755	3.783	2.582	3.188	4.353	3.496	4.287	5.811
9	2.133	2.649	3.641	2.454	3.031	4.143	3.242	3.971	5.389
10	2.065	2.568	3.532	2.355	2.911	3.981	3.048	3.739	5.075
11	2.012	2.503	3.444	2.275	2.815	3.852	2.897	3.557	4.828
12	1.966	2.448	3.371	2.210	2.736	3.747	2.773	3.410	4.633
13	1.928	2.403	3.310	2.155	2.670	3.659	2.677	3.290	4.472
14	1.895	2.363	3.257	2.108	2.614	3.585	2.592	3.189	4.336
15	1.866	2.329	3.212	2.068	2.566	3.520	2.521	3.102	4.224
16	1.842	2.299	3.172	2.032	2.523	3.463	2.458	3.028	4.124
17	1.820	2.272	3.136	2.001	2.486	3.415	2.405	2.962	4.038
18	1.800	2.249	3.106	1.974	2.453	3.370	2.357	2.906	3.961
19	1.781	2.228	3.078	1.949	2.423	3.331	2.315	2.855	3.893
20	1.765	2.208	3.052	1.926	2.396	3.295	2.275	2.807	3.832
21	1.750	2.190	3.028	1.905	2.371	3.262	2.241	2.768	3.776
22	1.736	2.174	3.007	1.887	2.350	3.233	2.208	2.729	3.727
23	1.724	2.159	2.987	1.869	2.329	3.206	2.179	2.693	3.680
24	1.712	2.145	2.969	1.853	2.309	3.181	2.154	2.663	3.638
25	1.702	2.132	2.952	1.838	2.292	3.158	2.129	2.632	3.601
30	1.657	2.080	2.884	1.778	2.220	3.064	2.029	2.516	3.446
35	1.623	2.041	2.833	1.732	2.166	2.994	1.957	2.431	3.334
40	1.598	2.010	2.793	1.697	2.126	2.941	1.902	2.365	3.250
45	1.577	1.986	2.762	1.669	2.092	2.897	1.857	2.313	3.181
50	1.560	1.965	2.735	1.646	2.065	2.863	1.821	2.296	3.124

Source: From D. C. Montgomery (1985), *Introduction to Statistical Quality Control*, used by permission of John Wiley & Sons, Inc.

TABLE A.6 Percentage Points of the *F*-Distribution: Upper 5 Percentage Points

Numerator Degrees of Freedom (ν_1)

		1	2	3	4	5	6	7	8	9	10	12	15	20	24	30	40	60	120	∞
	1	161.4	199.5	215.7	224.6	230.2	234.0	236.8	238.9	240.5	241.9	243.9	245.9	248.0	249.1	250.1	251.1	252.2	253.3	254.3
	2	18.51	19.00	19.16	19.25	19.30	19.33	19.35	19.37	19.38	19.41	19.40	19.43	19.45	19.45	19.46	19.47	19.48	19.49	19.50
	3	10.13	9.55	9.28	9.12	9.01	8.94	8.89	8.85	8.81	8.79	8.74	8.70	8.66	8.64	8.62	8.59	8.57	8.55	8.53
	4	7.71	6.94	6.59	6.39	6.26	6.16	6.09	6.04	6.00	5.96	5.91	5.86	5.80	5.77	5.75	5.72	5.69	5.66	5.63
	5	6.61	5.79	5.41	5.19	5.05	4.95	4.88	4.82	4.77	4.74	4.68	4.62	4.56	4.53	4.50	4.46	4.43	4.40	4.36
	6	5.99	5.14	4.76	4.53	4.39	4.28	4.21	4.15	4.10	4.06	4.00	3.94	3.87	3.84	3.81	3.77	3.74	3.70	3.67
	7	5.59	4.74	4.35	4.12	3.97	3.87	3.79	3.73	3.68	3.64	3.57	3.51	3.44	3.41	3.38	3.34	3.30	3.27	3.23
	8	5.32	4.46	4.07	3.84	3.69	3.58	3.50	3.44	3.39	3.35	3.28	3.22	3.15	3.12	3.08	3.04	3.01	2.97	2.93
	9	5.12	4.26	3.86	3.63	3.48	3.37	3.29	3.23	3.18	3.14	3.07	3.01	2.94	2.90	2.86	2.83	2.79	2.75	2.71
	10	4.96	4.10	3.71	3.48	3.33	3.22	3.14	3.07	3.02	2.98	2.91	2.85	2.77	2.74	2.70	2.66	2.62	2.58	2.54
	11	4.84	3.98	3.59	3.36	3.20	3.09	3.01	2.95	2.90	2.85	2.79	2.72	2.65	2.61	2.57	2.53	2.49	2.45	2.40
Denominator Degrees of Freedom (ν_2)	12	4.75	3.89	3.49	3.26	3.11	3.00	2.91	2.85	2.80	2.75	2.69	2.62	2.54	2.51	2.47	2.43	2.38	2.34	2.30
	13	4.67	3.81	3.41	3.18	3.03	2.92	2.83	2.77	2.71	2.67	2.60	2.53	2.46	2.42	2.38	2.34	2.30	2.25	2.21
	14	4.60	3.74	3.34	3.11	2.96	2.85	2.76	2.70	2.65	2.60	2.53	2.46	2.39	2.35	2.31	2.27	2.22	2.18	2.13
	15	4.54	3.68	3.29	3.06	2.90	2.79	2.71	2.64	2.59	2.54	2.48	2.40	2.33	2.29	2.25	2.20	2.16	2.11	2.07
	16	4.49	3.63	3.24	3.01	2.85	2.74	2.66	2.59	2.54	2.49	2.42	2.35	2.28	2.24	2.19	2.15	2.11	2.06	2.01
	17	4.45	3.59	3.20	2.96	2.81	2.70	2.61	2.55	2.49	2.45	2.38	2.31	2.23	2.19	2.15	2.10	2.06	2.01	1.96
	18	4.41	3.55	3.16	2.93	2.77	2.66	2.58	2.51	2.46	2.41	2.34	2.27	2.19	2.15	2.11	2.06	2.02	1.97	1.92
	19	4.38	3.52	3.13	2.90	2.74	2.63	2.54	2.48	2.42	2.38	2.31	2.23	2.16	2.11	2.07	2.03	1.98	1.93	1.88
	20	4.35	3.49	3.10	2.87	2.71	2.60	2.51	2.45	2.39	2.35	2.28	2.20	2.12	2.08	2.04	1.99	1.95	1.90	1.84
	21	4.32	3.47	3.07	2.84	2.68	2.57	2.49	2.42	2.37	2.32	2.25	2.18	2.10	2.05	2.01	1.96	1.92	1.87	1.81
	22	4.30	3.44	3.05	2.82	2.66	2.55	2.46	2.40	2.34	2.30	2.23	2.15	2.07	2.03	1.98	1.94	1.89	1.84	1.78
	23	4.28	3.42	3.03	2.80	2.64	2.53	2.44	2.37	2.32	2.27	2.20	2.13	2.05	2.01	1.96	1.91	1.86	1.81	1.76
	24	4.26	3.40	3.01	2.78	2.62	2.51	2.42	2.36	2.30	2.25	2.18	2.11	2.03	1.98	1.94	1.89	1.84	1.79	1.73
	25	4.24	3.39	2.99	2.76	2.60	2.49	2.40	2.34	2.28	2.24	2.16	2.09	2.01	1.96	1.92	1.87	1.82	1.77	1.71
	26	4.23	3.37	2.98	2.74	2.59	2.47	2.39	2.32	2.27	2.22	2.15	2.07	1.99	1.95	1.90	1.85	1.80	1.75	1.69
	27	4.21	3.35	2.96	2.73	2.57	2.46	2.37	2.31	2.25	2.20	2.13	2.06	1.97	1.93	1.88	1.84	1.79	1.73	1.67
	28	4.20	3.34	2.95	2.71	2.56	2.45	2.36	2.29	2.24	2.19	2.12	2.04	1.96	1.91	1.87	1.82	1.77	1.71	1.65
	29	4.18	3.33	2.93	2.70	2.55	2.43	2.35	2.28	2.22	2.18	2.10	2.03	1.94	1.90	1.85	1.81	1.75	1.70	1.64
	30	4.17	3.32	2.92	2.69	2.53	2.42	2.33	2.27	2.21	2.16	2.09	2.01	1.93	1.89	1.84	1.79	1.74	1.68	1.62
	40	4.08	3.23	2.84	2.61	2.45	2.34	2.25	2.18	2.12	2.08	2.00	1.92	1.84	1.79	1.74	1.69	1.64	1.58	1.51
	60	4.00	3.15	2.76	2.53	2.37	2.25	2.17	2.10	2.04	1.99	1.92	1.84	1.75	1.70	1.65	1.59	1.53	1.47	1.39
	120	3.92	3.07	2.68	2.45	2.29	2.17	2.09	2.02	1.96	1.91	1.83	1.75	1.66	1.61	1.55	1.50	1.43	1.35	1.25
	∞	3.84	3.00	2.60	2.37	2.21	2.10	2.01	1.94	1.88	1.83	1.75	1.67	1.57	1.52	1.46	1.39	1.32	1.22	1.00

Source: From M. Merrington and C. M. Thompson (1943), Tables of Percentage Points of the Inverted Beta (F) Distribution, *Biometrika*, used by permission of Oxford University Press.

TABLE A.7   Percentage Points of the *F*-Distribution: Upper 2.5 Percentage Points

		Numerator Degrees of Freedom (ν_1)																	
	1	2	3	4	5	6	7	8	9	10	12	15	20	24	30	40	60	120	∞
1	647.8	799.5	864.2	899.6	921.8	937.1	948.2	956.7	963.3	968.6	976.7	984.9	993.1	997.2	1001.0	1006.0	1010.0	1014.0	1018.0
2	38.51	39.00	39.17	39.25	39.30	39.33	39.36	39.37	39.39	39.40	39.41	39.43	39.45	39.46	39.46	39.47	39.48	39.49	39.50
3	17.44	16.04	15.44	15.10	14.88	14.73	14.62	14.54	14.47	14.42	14.34	14.25	14.17	14.12	14.08	14.04	13.99	13.95	13.90
4	12.22	10.65	9.98	9.60	9.36	9.20	9.07	8.98	8.90	8.84	8.75	8.66	8.56	8.51	8.46	8.41	8.36	8.31	8.26
5	10.01	8.43	7.76	7.39	7.15	6.98	6.85	6.76	6.68	6.62	6.52	6.43	6.33	6.28	6.23	6.18	6.12	6.07	6.02
6	8.81	7.26	6.60	6.23	5.99	5.82	5.70	5.60	5.52	5.46	5.37	5.27	5.17	5.12	5.07	5.01	4.96	4.90	4.85
7	8.07	6.54	5.89	5.52	5.29	5.12	4.99	4.90	4.82	4.76	4.67	4.57	4.47	4.42	4.36	4.31	4.25	4.20	4.14
8	7.57	6.06	5.42	5.05	4.82	4.65	4.53	4.43	4.36	4.30	4.20	4.10	4.00	3.95	3.89	3.84	3.78	3.73	3.67
9	7.21	5.71	5.08	4.72	4.48	4.32	4.20	4.10	4.03	3.96	3.87	3.77	3.67	3.61	3.56	3.51	3.45	3.39	3.33
10	6.94	5.46	4.83	4.47	4.24	4.07	3.95	3.85	3.78	3.72	3.62	3.52	3.42	3.37	3.31	3.26	3.20	3.14	3.08
11	6.72	5.26	4.63	4.28	4.04	3.88	3.76	3.66	3.59	3.53	3.43	3.33	3.23	3.17	3.12	3.06	3.00	2.94	2.88
12	6.55	5.10	4.47	4.12	3.89	3.73	3.61	3.51	3.44	3.37	3.28	3.18	3.07	3.02	2.96	2.91	2.85	2.79	2.72
13	6.41	4.97	4.35	4.00	3.77	3.60	3.48	3.39	3.31	3.25	3.15	3.05	2.95	2.89	2.84	2.78	2.72	2.66	2.60
14	6.30	4.86	4.24	3.89	3.66	3.50	3.38	3.29	3.21	3.15	3.05	2.95	2.84	2.79	2.73	2.67	2.61	2.55	2.49
15	6.20	4.77	4.15	3.80	3.58	3.41	3.29	3.20	3.12	3.06	2.96	2.86	2.76	2.70	2.64	2.59	2.52	2.46	2.40
16	6.12	4.69	4.08	3.73	3.50	3.34	3.22	3.12	3.05	2.99	2.89	2.79	2.68	2.63	2.57	2.51	2.45	2.38	2.32
17	6.04	4.62	4.01	3.66	3.44	3.28	3.16	3.06	2.98	2.92	2.82	2.72	2.62	2.56	2.50	2.44	2.38	2.32	2.25
18	5.98	4.56	3.95	3.61	3.38	3.22	3.10	3.01	2.93	2.87	2.77	2.67	2.56	2.50	2.44	2.38	2.32	2.26	2.19
19	5.92	4.51	3.90	3.56	3.33	3.17	3.05	2.96	2.88	2.82	2.72	2.62	2.51	2.45	2.39	2.33	2.27	2.20	2.13
20	5.87	4.46	3.86	3.51	3.29	3.13	3.01	2.91	2.84	2.77	2.68	2.57	2.46	2.41	2.35	2.29	2.22	2.16	2.09
21	5.83	4.42	3.82	3.48	3.25	3.09	2.97	2.87	2.80	2.73	2.64	2.53	2.42	2.37	2.31	2.25	2.18	2.11	2.04
22	5.79	4.38	3.78	3.44	3.22	3.05	2.93	2.84	2.76	2.70	2.60	2.50	2.39	2.33	2.27	2.21	2.14	2.08	2.00
23	5.75	4.35	3.75	3.41	3.18	3.02	2.90	2.81	2.73	2.67	2.57	2.47	2.36	2.30	2.24	2.18	2.11	2.04	1.97
24	5.72	4.32	3.72	3.38	3.15	2.99	2.87	2.78	2.70	2.64	2.54	2.44	2.33	2.27	2.21	2.15	2.08	2.01	1.94
25	5.69	4.29	3.69	3.35	3.13	2.97	2.85	2.75	2.68	2.61	2.51	2.41	2.30	2.24	2.18	2.12	2.05	1.98	1.91
26	5.66	4.27	3.67	3.33	3.10	2.94	2.82	2.73	2.65	2.59	2.49	2.39	2.28	2.22	2.16	2.09	2.03	1.95	1.88
27	5.63	4.24	3.65	3.31	3.08	2.92	2.80	2.71	2.63	2.57	2.47	2.36	2.25	2.19	2.13	2.07	2.00	1.93	1.85
28	5.61	4.22	3.63	3.29	3.06	2.90	2.78	2.69	2.61	2.55	2.45	2.34	2.23	2.17	2.11	2.05	1.98	1.91	1.83
29	5.59	4.20	3.61	3.27	3.04	2.88	2.76	2.67	2.59	2.53	2.43	2.32	2.21	2.15	2.09	2.03	1.96	1.89	1.81
30	5.57	4.18	3.59	3.25	3.03	2.87	2.75	2.65	2.57	2.51	2.41	2.31	2.20	2.14	2.07	2.01	1.94	1.87	1.79
40	5.42	4.05	3.46	3.13	2.90	2.74	2.62	2.53	2.45	2.39	2.29	2.18	2.07	2.01	1.94	1.88	1.80	1.72	1.64
60	5.29	3.93	3.34	3.01	2.79	2.63	2.51	2.41	2.33	2.27	2.17	2.06	1.94	1.88	1.82	1.74	1.67	1.58	1.48
120	5.15	3.80	3.23	2.89	2.67	2.52	2.39	2.30	2.22	2.16	2.05	1.94	1.82	1.76	1.69	1.61	1.53	1.43	1.31
∞	5.02	3.69	3.12	2.79	2.57	2.41	2.29	2.19	2.11	2.05	1.94	1.83	1.71	1.64	1.57	1.48	1.39	1.27	1.00

Denominator Degrees of Freedom (ν_2)

Source: From M. Merrington and C. M. Thompson (1943), Tables of Percentage Points of the Inverted Beta (F) Distribution, *Biometrika*, used by permission of Oxford University Press.

REFERENCES

ANSI/ASQC Z1.4 (1981) *Sampling Procedures and Tables for Inspection by Attributes*. American Society for Quality Control, Milwaukee, WI.

ANSI/ASQC Z1.9 (1981) *Sampling Procedures and Tables for Inspection by Variables for Percent Non-Conforming*. American Society for Quality Control, Milwaukee, WI.

Box, G. E. P. and Draper, N. R. (1987) *Empirical Model Building and Response Surfaces*. Wiley, New York.

Box, G. E. P. and Tyssedal, J. (1996) Projective properties of certain orthogonal arrays. *Biometrika* 83(4), 950–955.

Box, G. E. P., Hunter, W. G., and Hunter, J. S. (1978) *Statistics for Experimenters*. Wiley, New York.

Cornell, A. C. (2002) *Experiments with Mixtures: Designs, Models, and the Analysis of Mixture Data*. Wiley, New York.

Grant, E. L. and Leavenworth, R. S. (1980) *Statistical Quality Control*, McGraw-Hill, New York.

Ibbotson, R. G. (2006) *Stocks, Bonds, Bills and Inflation, 2006 Yearbook*. Ibbotson Associates, Chicago, IL.

Joglekar, A. M. (2003) *Statistical Methods for Six Sigma in R&D and Manufacturing*. Wiley, New York.

Joglekar, A. M., Joshi, N., Song, Y., and Ergun, J. (2007) Mathematical model to predict coat weight variability in a pan coating process. *Pharm. Dev. Technol.* 1(3), 297–306.

Neter, J., Wasserman, W., and Kutner, M. H. (1990) *Applied Linear Statistical Models*. Irwin, Boston, MA.

Phadke, M. S. (1989) *Quality Engineering Using Robust Design*. Prentice Hall, Englewood Cliffs, NJ.

Software (2009) JMP by SAS Institute Inc., Statgraphics by Stat Point Inc., Design-Expert by Stat-Ease, Inc.

Steinberg, D. M. and Bursztyn, D. (1994) Dispersion effects in robust-design experiments with noise factors. *Journal of Quality Technology* 26(1), 12–20.

Taguchi, G. (1987) *System of Experimental Design*. Vols. 1 and 2. UNIPUB, Kraus International Publications, White Plains, New York.

Taguchi, G., Elsayed, E. A., and Hsiang, T. (1989) *Quality Engineering in Production Systems*. McGraw-Hill, New York.

Ye, C., Liu, J., Ren, F. and Okafo, N. (2000) Design of experiment and data analysis by JMP® (SAS institute) in analytical method validation, *J Pharm. Biomed. Anal.* 23, 581–589.

INDEX

Industrial Statistics: Practical Methods and Guidance for Improved Performance   By Anand M. Joglekar
Copyright © 2010 John Wiley & Sons, Inc.